A GUIDE TO INTEGRATED WARM WATER AQUACULTURE

A GUIDE TO INTEGRATED WARM WATER AQUACULTURE

by

DAVID LITTLE

and

JAMES MUIR

INSTITUTE OF AQUACULTURE PUBLICATIONS
UNIVERSITY OF STIRLING

Published by the Institute of Aquaculture,
University of Stirling,
Stirling FK9 4LA,
Scotland.

First published 1987

British Library
Cataloguing in Publication Data
Little, David
A Guide to Integrated Warm Water Aquaculture
1. Fish-culture
I. Title II. Muir, James
III. Institute of Aquaculture
639.3 SH151

ISBN 0-901636-71-1

Printed in Great Britain
by W. M. Bett Ltd, Tillicoultry

CONTENTS

FIGURES

TABLES

BOXES

PREFACE

In the course of our research and teaching work in aquaculture, we have become increasingly convinced of the important role to be played in integrating aquaculture with other activities. Not only are useful resources preserved within rural communities, but aquaculture may itself be a crucial element in the successful handling of difficult materials.

While some of the practices of 'integration' are as old as the oldest traditions of aquaculture, there are many new developments, new crops, processes, and wastes, and many areas where a carefully-chosen combination of resources could be effective. This guide has been written in order to try to collect some of the information which is now available, and we hope, to analyse this information to the level at which users of the book will be able to apply techniques, analyse results, and extend and improve methodology. We would judge our intentions to have been successful if we receive a generous supply of comments, news of field experience, and updated field data from our readers!

AIT, Bangkok
Stirling

DAVID LITTLE
JAMES MUIR

ACKNOWLEDGEMENTS

We have borrowed heavily of time, wisdom, intellectual effort, inspiration, patience, and sense of humour; to the many people who have generously contributed to our efforts we extend our heartfelt thanks. We would particularly like to acknowledge Jorge Fuchs and Albert Tacon for their contributions in providing data on waste materials and integrated farming systems, and Peter Edwards of AIT for access to valuable unpublished material and for his inspiration to both of us – we're sure that he'll produce the definitive version sometime!

Our gratitude goes also to Malcolm Beveridge, Roy Clarke, Hilary Duggua and Mike Phillips of Stirling, and to Steve Hawkins of the Zoology Department of Manchester University for their excellent assistance and advice throughout our project.

Thanks are particularly due to our 'production department' Maggie Beveridge and Jan Candy for their skilful help with editing the text and turning it into English, and to Kanchana Taechlertpaisarn, Laura Cummings, and Moira Stewart for deciphering it in the first place. Lyn North very capably turned our crude sketchings into the excellent illustrations now gracing the text.

Our thanks also go to Professor R. J. Roberts, of the Institute of Aquaculture for his support of our work, to our other colleagues in Stirling and Bangkok for their unstinting cooperation and encouragement, and to the Natural Environmental Research Council and the Overseas Development Administration of the UK, and the EEC Development Programme for their financial support at various stages in our work in aquaculture.

Finally any errors, whether of fact or interpretation, we acknowledge as our own.

INTRODUCTION

In recent years a new addition to food production has appeared in many parts of the world. Attention is turning towards fish farming, 'aquaculture', as a food source.

The exotic nature of farming fish is reflected in its varying position within the wider farming system. Sometimes it has developed quite independently from mainstream agriculture, or, in contrast, has become integrated to some extent with the production of livestock and crops. The culture of various carnivorous fish for instance, is long established in some places, and has developed into a specialised industry quite separate from other types of farming. The origins of this type of aquaculture are often to be found in capture fisheries or sporting interests. However, where farmed fish have been economically and nutritionally most important, integration within the farm has often been a practical necessity. 'Integrated' fish culture probably first developed in densely populated parts of Asia and Central Europe, which were essentially agriculturally based and had limited access to wild fish stocks. Although these areas might have been open to imports of feed and fertiliser, there would certainly have been practical limits to wider or more intensive use of them. The development of ways and means to produce fish was therefore inevitably directed towards using resources within the farm or from the immediate area.

The social and economic conditions that encouraged integrated fish culture to evolve in the past now occur in many other regions of the world. Integrated aquaculture can compliment and improve the overall efficiency of many types of farm. For instance, the more efficient use of water and labour, as well as waste recycling into fish are two obvious examples. Conversely, livestock and arable crop production may be the only source of feeds and fertilisers available at low enough cost to make fish culture possible.

The most common forms of integration are those where there is a direct and simple link between activities; such as the use of animal or crop wastes as fish feed or fertiliser. The use of irrigation structures and water supplies to raise fish, or the production of fish in rice fields are also examples.

Less commonly, some form of processing is used to make wastes and by-products more useful to fish culture. Here the links between the integrated 'components' become less direct, but often easier to manage. Drying and decomposition of bulk waste materials to a reduced and more stable state is a typical example. More complicated methods, involving fermentation or decomposition, can be used to add value or usefulness to waste. Anaerobic digestors, for instance, are now increasingly important in India and China for the production of biogas and other products such as fertile slurries and sludge that can be used in fish culture. Another route through which wastes can be upgraded is their use as a substrate for growing intermediate organisms, which are subsequently fed to fish. Worms, insects and trash fish are being utilised in this way.

Aquaculture production may also produce 'resources' that benefit other activities on the integrated farm; fish processing wastes can be recycled directly as a feed for livestock, pond water or sediments can be used for arable crop production, and herbivorous fish can keep irrigation channels clear of weed.

The potential of integrated aquaculture

The concept of integrating fish production with other activities is not new. Traditional Chinese and European aquaculture has relied almost exclusively on the resources shared and recycled locally with animal and crop production. This is in contrast to the relatively recent development of intensive fish culture with its emphasis on a more complex technology and high inputs of resources and energy. This development parallels a worldwide trend towards separation and specialisation in food production, which is often justified as a more productive and 'efficient' approach. While yields are increased in terms of area and labour input, efficiency in terms of resource use is frequently extremely low.

Increased production from aquaculture may be of key importance in many developing countries, especially where fish consumption is high, population growth is significant, and natural fisheries have been depleted. But is an integrated approach a viable alternative given the apparent trend to more intensive methods?

On a practical level, extensive aquaculture is the most common form of production, using traditional techniques to yield a low production of fish per unit area from minimum inputs (see Fig. 0.1). A lack of inputs is often cited for the continued use of such systems in many developing countries. Whilst extensive methods will have a continuing role to play in the energy efficient production of fish, their low yields mean that they are becoming increasingly uncompetitive with other forms of land and water use. Integration with agriculture or other activities may improve this position by creating a more productive system. Compared with the simple technology of extensive fish culture and the expensive higher technology of many intensive farms, integrated aquaculture may be a viable and often superior intermediate option, giving relatively high yields with fairly low input cost. Further evidence for this view can be found in the rapid expansion of integrated fish farming in many parts of the world. In Taiwan, for instance, it is estimated that over a third of all fish ponds are integrated in some way, and that such systems are continuing to grow in popularity (Crocker, 1983). However, the traditional practice of combined fish and rice culture, which is a classic method of integrating land and water use, is declining in many regions due to incompatability with modern methods of rice culture. Thus, integration of fish culture within the farm is not, nor has it ever been, static. The wider farming system changes with new technical developments and market forces. Opportunities for the integration of activities inevitably rise and fall alongside these technical and market changes.

Rationale for development of integrated aquaculture

Integrated aquaculture has become commonly associated with the profitable use of animal and crop wastes. Where the amount of waste is small, the production of fish is normally the most important goal. However, when wastes are produced on a larger scale or in more concentrated form, their cheap and efficient disposal may be more significant.

As intensive farming and urbanisation have increased, the problems of disposal of such concentrated wastes have become particularly acute. 'Dumping' of wastes, whether on economic, energy or environmental grounds, is becoming increasingly undesirable while conventional forms of waste treatment are often too expensive for use in many locations. Fish culture may then become an alternative and profitable means of disposing of such wastes.

The energy crisis of the early 1970s gave particular impetus to the critical assessment of traditional integrated systems. The success of such methods in China has prompted interest by many developing countries. Integrated aquaculture has been recognised as having potential for improving the position of the rural poor. Some critics are less optimistic, but the approach may at least have a role in providing high protein fish in a simple and flexible way, without necessarily requiring additional resources outside the existing system. However, the possibilities of a simple technology transfer of integrated techniques may not be immediately realised. China has not only been culturing fish for longer than most nations, but it also has evolved a form of integration suitable for the specific social, economic and cultural conditions of that country and its people. While the principles are important and valuable in guiding development elsewhere, direct transplantation of techniques will require careful consideration of local resources, conditions, and attitudes, and must be considered a gradual process. Certainly, attempts to adapt integrated aquaculture methods to present traditions and farming systems will be more successful than those that are based on radical change.

The 'commonsense' of integration is often self-evident in places with any history of subsistence fish culture. Fish ponds will perhaps not only serve to raise fish but will also supply water for surrounding vegetables and fruit, sometimes even domestic purposes, and so the return of wastes from these activities may follow on naturally. The extension of integrated farming technology into a particular area requires that it must be not only socially acceptable but also be capable of uptake by farmers with the minimum of risk. The demonstration of improvements in economic terms would also of course accelerate acceptance by farmers.

About this book

This book has the aim of describing the principles of integrated aquaculture and illustrating the means by which such farming can be developed successfully.

It is our aim to concentrate on tropical aquaculture, based generally on pond systems; the more intensive forms of integration such as use of power station effluents will not be considered. It is hoped that the information provided will be available as a reference and guide to those wishing to improve or develop their own integrated systems.

This book should however be used with the knowledge that integrated systems are highly site specific and involve many variables. A direct comparison between systems may not always be useful, and even well planned and monitored trials will produce results that are not directly applicable elsewhere. We have deliberately avoided exhaustive quoting of scientific or technical references; there should however be sufficient of the main sources available for those specifically interested; the main aim is to provide explanation and practical guidance to a general readership. We do nevertheless hope to provide the basic 'building blocks' with which an integrated system can be developed—the 'boxes' in which are explained some of the calculations involved in selecting and designing integrated systems should be particularly useful.

As techniques evolve, and the process of 'tailoring' integrated farming to available conditions becomes better understood, so the quantification of methods will become more definable. We would therefore urge the intending user of the techniques we describe to explore changes and improvements to these basics, and to participate in the wider development of understanding in this area.

The sequence of the text is as follows. We first discuss the environment surrounding and supporting the mechanism of fish production in a waste-fed pond as this is the basis on which most integrated systems depend. Later we consider the concept of fish production as a single component within a more complex integrated farming system. We consider in turn, livestock production, integrated crops, human waste and processed products. Attention is directed throughout to integration of aquaculture at the level of the individual farm or village as it is here that such practices can reveal real benefits. As many technical terms and fish species names are used throughout, a glossary is provided to guide those unfamiliar with particular names or descriptions.

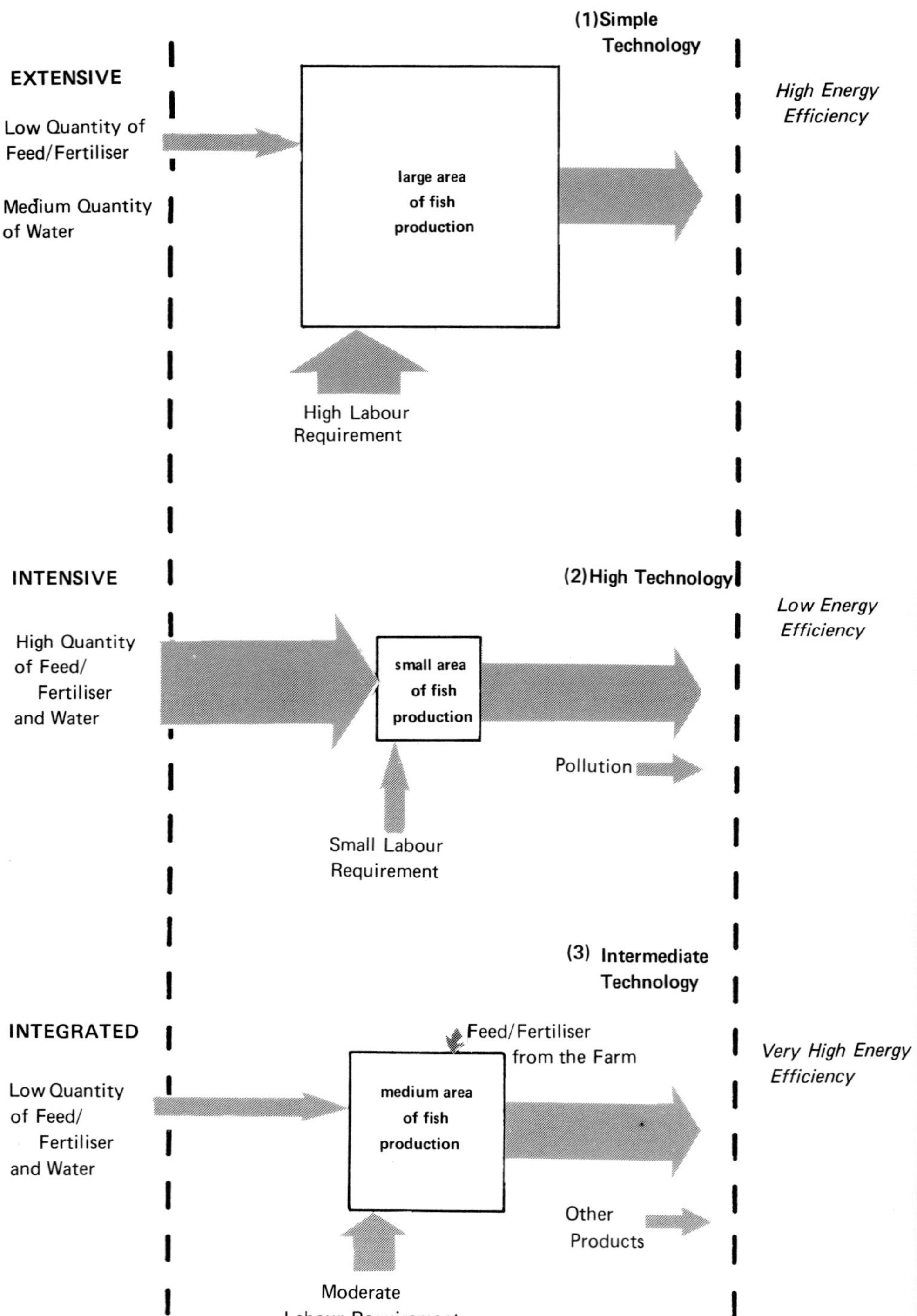

Fig. 0.1. Intensive, extensive and integrated aquaculture.

CHAPTER 1
THE FISH POND ENVIRONMENT

1.1 Introduction

Most fish are still cultured in simple ponds with limited, if any, water exchange. The great variety in fish yields reported for such systems (50-13000 kg/ha/yr) reflects the wide variation in species employed, local conditions and levels of management. The limited water exchange allows a build up of the nutrients supporting the growth of natural feeds, especially if fertilisers, manures or other waste products are added. Thus, natural feeds in the pond may reach levels far in excess of those found in most natural aquatic systems. Since the density of fish that can be grown in a volume of water is, within limits, directly related to the abundance of food, this can improve yields without the use of costly feed inputs. The stimulation of natural feeds can be based on almost any nutrient source, whether used singly or in combination with others. The system is thus extremely flexible and is ideal for accepting the range of raw materials available in integrated farming. The basic principle of pond feeding is that nutrients dissolved in water and sunlight energy are converted into usable cell material for consumption by the fish.

Green, eutrophic, water is typical of fertilised warm water fish ponds. This is often the simplest and most obvious sign that the natural web of food organisms within the pond is productive, and that yields of stocked fish will usually be higher than in a natural system. Indeed, such is the quality of this

Box 1.1. What yields can be obtained in different systems?

			kg/ha/yr
(1)	Unfertilised, poorly managed systems	:	50–200
(2)	Fertilised, waste fed systems, with stock management	:	5000–10000
(3)	Fertilised and supplementary-fed pond systems	:	5000–15000
(4)	More complete feeding, more water exchange, supplementary aeration	:	8000–40000
(5)	Cage culture or flowing water pond systems, with complete feeding and water exchange	:	200–2000 ($\times$ 10^3)

natural feed that fish yields may reach or exceed those possible with conventional fish feeds.

The algae, or phytoplankton, are the primary and most obvious natural food in a fertilised pond. They basically fix nutrients in the water in a form available as food for plankton-eating fish by means of photosynthesis. Production is thus related to both nutrient and sunlight energy input; tropical ponds, receiving high levels of sunlight, have the greatest potential for production of natural feeds. These in turn form the basis of a balance of animal and plant communities, forming a web of feeding relationships. This solar-powered food production is often termed the 'autotrophic' pathway. Phytoplankton are also essential in producing oxygen in the pond. A second fundamental system, the 'heterotrophic' pathway, is based on the breakdown of dead plant and animal material in the pond by bacteria. Up to certain limits, increases in nutrient input results in an increase in phytoplankton, bacteria and zooplankton production, allowing a consequent increase in organisms feeding on them. Figure 1.1 shows the typical feeding web within the fertilised pond.

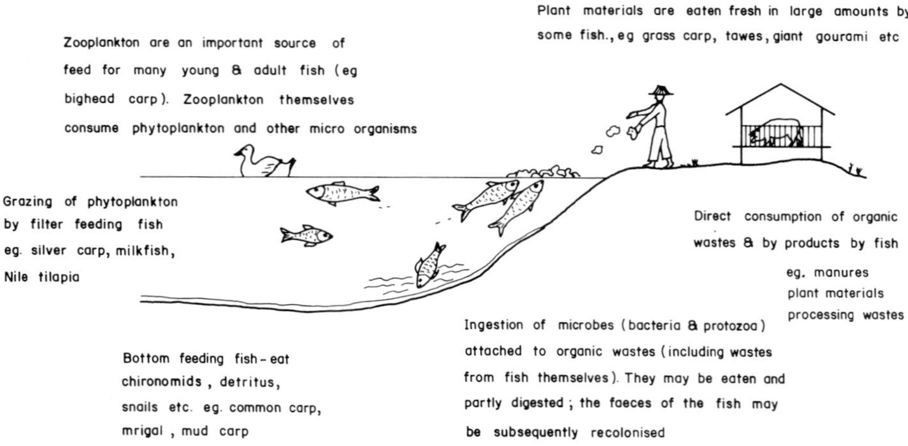

Fig. 1.1. Feeding pathways in the waste-fed fish pond.

The relative qualities and types of food organism within the pond define the potential for fish yield. Actual yield depends further on the type and numbers of fish stocked, as different species feed on different feed organisms, ranging from phytoplankton to macrophytes, zooplankton, bacteria and detritus. In most cases the basis for fish yield is the light-limited 'autotrophic' phytoplankton production, though in heavily manured ponds the heterotrophic use of wastes by bacteria and protozoa may also be important. In all cases, the primary goal of successful pond operation is the managing of the different pond organisms to maximise the opportunities for fish feeding and growth, and to maximise the absorption of waste materials and their conversion into more useful and stable products.

Box 1.2. Relation between daily yield and annual production.

An efficient waste-fed pond can support growth of 30 kg/ha/day. If this were available year round, potential production would be 30 × 365 = 10950 kg/ha/yr!

In practice, many locations have perhaps 200–250 growing days per year, because of availability of fry or water, climatic changes, and the need to drain and harvest ponds.

Thus 30 × 200 = 6000 kg/ha/yr

which is still an excellent output.

The term 'sunlit rumen' has been coined (Schroeder, 1978) to describe the warm water manured fish pond and the interaction of processes occurring within it. The maintenance of the 'dynamic balance' of organisms in the pond is usually facilitated by the stocking of a variety of fish species enabling fullest exploitation of the pond's space and its natural diverse food web. In practice, most of the fish stocked should be those capable of feeding low in the food chain on the natural feeds described, rather than predatory or carnivorous fish. The process of transfer of nutrients through this food web is described in Figure 1.2.

Fish require adequate oxygen as well as food for healthy and sustained growth. If the pond is correctly managed, sufficient oxygen should be available to sustain the metabolic demands of the food web, including the fish. It is essential for the effective management of ponds, and the efficient utilisation of the resources within integrated systems, to understand how the food and nutrient cycles of the pond operate and interrelate. These are therefore discussed in more detail in the following sections.

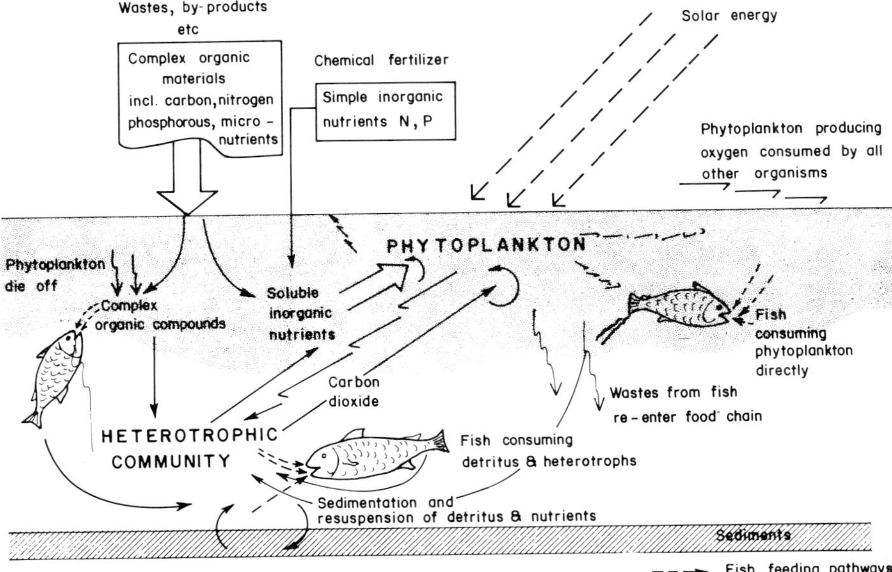

Fig. 1.2. Nutrient and oxygen transfer in the waste-fed fish pond.

1.2 Action of wastes on fish ponds

When waste materials enter a pond system, they may either be consumed directly by the fish, or may act indirectly after being broken down to support the pond food web. The relative importance of these routes is the subject of some controversy. Present evidence suggests that the value of the manure as direct feed for fish is small compared to its beneficial effect on natural food production within the pond ecosystem.

1.2.1 Wastes as direct feeds

In general, relatively poor responses have been found in growth trials using animal manures as direct feed (Lu and Kevern, 1975). Although some workers e.g. Spaturu (1976, 1977) have found little evidence of direct consumption of manures by fish, this undoubtedly does occur in such species as the silver striped catfish *Pangasius sutchi* and the Nile tilapia (*Oreochromis niloticus*). With manures containing large amounts of partially digested feed, direct ingestion of unconsumed elements by the fish may be the most important source of feeding. Intestinal bacteria are present in all animal manures and these may also attract ingestion by fish. Kerns and Roelofs (1977) found dried poultry waste to be of low nutritional value to fish but considered that the bacteria and protozoa contained in these particles were of high nutritional value to the fish. However, some fish such as the Nile tilapia seem capable of using low quality waste or even non-living detritus as a direct energy source (Edwards, 1982[a]). Recycling of animal wastes to fish and other animals has been the subject of much research, as reviewed by Muller (1982).

1.2.2 Wastes as indirect feeds

Waste could potentially be made available indirectly by either the autotrophic or heterotrophic pathways. The limits to fish production from the autotrophic, plant-based feeding pathway depends largely on the rate of plant production possible within the fish pond, the so called primary production. However, there is no simple relationship

$$\text{nutrient input} + \text{solar energy} = \text{fish production.}$$

The absorption and use of nutrients, by any species of phytoplankton, is affected by temperature and light intensity. In addition, continual cropping of these algae by planktivorous fish make for a dynamic situation since this removal results in more light and nutrients being made available for the rest of the algae (Reich, 1975). The reduction in 'autoshading' made possible by stocking planktivorous fish has been found to increase production of the algae and the fish, both in terms of pond area and fertiliser used.

The relative merits of organic and inorganic fertilisers and their effectiveness in auto- and heterotrophic food chains has caused some controversy (Colman and Edwards, 1985). Organic wastes are undoubtedly valuable sources of nutrients to power 'autotrophic' production and as substrates for the heterotrophic community; they contain not only nitrogen, phosphorus and micronutrients, but also large quantities of carbon. Studies have indicated that inorganic fertilisers alone produce lower fish yields (10-15 kg/ha/day) than is possible using organic wastes (~ 30 kg/ha/day) (Schroeder, 1978). The source of this improved production appears to be mainly from the stimulation of the autotrophic food chain; organic wastes are a more complete source of nutrients for dense blooms of phytoplankton and allow higher productivity than is possible with chemical fertilisers alone. The heterotrophic food chain is also stimulated both from the detritus in the waste and that produced within the pond in the form of dead and dying algae.

Box 1.3. How does food production relate to fish production?

A 10% conversion efficiency is commonly used for estimation.

Thus, 50 kg/day of phytoplankton should produce about 50 × 10% = 5 kg/day of fish if fed directly. In a longer food chain –

50 kg/day phytoplankton → 5 kg/day zooplankton
5 kg/day zooplankton → 0·5 kg/day fish

Note: These are useful estimates, but are not exact. Conversion efficiences vary considerably depending on nutritional value of feeds, type of fish, etc.

1.3 Autotrophic pathways

1.3.1 Phytoplankton

The importance of phytoplankton to the nutrition of pond fish, even those assumed to be totally dependent on them as a source of food, has been challenged on the grounds of—

(a) availability

The majority of phytoplankton in waste-fed ponds are very small (nannoplankton, less than 20 μm), but certain fish are adapted to remove them. Physical straining or filtering together with mucus entrapment seems to be the basis for the removal of the tiny algae. The silver carp (*Hypophthalmichthys molitrix*) has fine gill rakers and effectively removes particles, whether they be algae or bacterial aggregates, of less than 20 μm. Mucus secretion from the gill surface and other specialised structures appears to entrap food particles and the gill rakers prevent the resultant slime from escaping in the outward current (Colman and Edwards, 1985). Fine teeth on the pharyngeal bones are also reported to improve filtration efficiency in plankton-feeding tilapias (Bowen, 1982). Plankton availability and feeding is possibly cyclical in warm water pond systems. Thus, autoflocculation of single celled phytoplankton in which cells 'clump' together and daily partial die-off are possible mechanisms for increasing the size of the particle, and thus the usefulness of phytoplankton to fish.

(b) digestibility

Assuming that ingestion of large quantities of nannoplankton is possible, and this is supported by gut content analysis of planktivorous fish, the question of their digestibility must be considered. Observations on the presence of undigested blue-green algae, coccoid green algae and euglenoids in the faeces of fish have led to the common belief that these common and important algae were not utilised by the fish. Research on tilapia species (Moriarty and Moriarty, 1973) has shown that the digestion of blue-green algae (Cyanobacteria) increases throughout the day, as acid secretion increases and stomach pH falls (less than 2). However, the mechanism of digestion in other important cultured fish e.g. silver carp, remains unexplained. The physiological condition of the filtered algae may also affect digestibility. Sedimented and dead algae, essentially in a 'detrital state' and partly decomposed, may be better assimilated than living algae (Bitterlich, 1985).

(c) toxicity

Several phytoplankton species are known to be toxic to other organisms. Of the two groups of algae reported to cause fish mortalities, blue-green algae are potentially the most important in waste-fed ponds. Dinoflagellates causing so called 'red tides' are normally only found in coastal marine waters. It is suggested that *Microcystis* sp., a common algal component of fish diets in waste-fed ponds, may be toxic under certain circumstances (Ingram and Prescott, 1954), but evidence is contradictory. Most problems involving *Microcystis* sp. poisoning in the literature involved livestock drinking bloom-infested waters (Colman and Edwards, 1985). Recently, tests using *Microcystis* sp. toxic to mice produced no significant mortality when filtered and ingested by the Nile tilapia (M. Phillips, personal communication). Many reports related to fish deaths are most likely explained by anoxia or the effect of algal decomposition products.

In Israel a yellow-brown algae, *Prymnesium parvum,* has been shown to have affected fish culture by secretion of an extracellular substance (Sarig, 1971). However, it was only in brackish water that the algae caused problems. Regular monitoring is practised to avoid fish deaths and populations of the problem algae are controlled by chemical means.

1.3.2 Aquatic macrophytes

Submerged aquatic macrophytes rarely contribute to primary production in well-managed waste-fed ponds, since shading by dense phytoplankton prevents their growth. Floating aquatic macrophytes are generally discouraged in fish ponds because of adverse effects on fish production. They are thus of relatively little importance in fish production, though if available, e.g. from water courses, can be used as a basis for feeding (see section 5.3).

1.4 Heterotrophic pathways

1.4.1 Bacteria and protozoa

The colonisation by bacteria of wastes, whether of animal, plant or fish origin, plays an important role in the pond's food production. In theory this form of heterotrophic production is limited only by the availability of organic substrate and dissolved oxygen to support the process. Schroeder's (1978) illustration of the increased protein value of chopped straw and cotton strips on which pond bacteria were encouraged to grow suggests the importance of this pathway in production of high quality fish food. An increase in the protein contained in aggregate detritus from 6% to 24% was found by Odum and De la Cruz (1967). This was attributed to the development of a micro-ecosystem containing bacteria, protozoa and micro-algae. Several authors, quoted by Schroeder (1980), concluded that the microbial community provided essentially all the nutritional needs of the fish. It may also be possible that the detrital substrate, typically a collection of waste material particles and grazing micro-organisms, is cycled through the fish being repeatedly ingested, stripped of nutritious micro-organisms and once again voided for re-colonisation. Different particle sizes can be utilised within the system; it was suggested that aggregates from 6-20 μm can be trapped best by zooplankton, from 21-26 μm by silver carp and these greater than 16 μm by the bighead carp (*Aristichthys nobilis*) (Edwards, 1982[b]), though other species in a pond ecosystem are also likely to be able to handle these large particles.

1.4.2 Zooplankton and chironomids

Zooplankton and chironomids often fulfill a major role in the food web supporting fish populations in natural ecosystems, but occur as relatively low populations in densely stocked fish ponds. Investigation of gut contents may fail to substantiate the importance of these organisms in the diet of cultured fish as, in contrast to the less digestible phytoplankton, zooplankton are typically digested quite rapidly (Bitterlich and Graiger, 1984).

The scarcity of these larger organisms in waste-fed ponds may be due to several reasons. Firstly, there is evidence that zooplankton may be suppressed by the dense phytoplankton communities, normal in such ponds. Inhibitory secretions of certain algae may also be significant and elevation of pH (often exceeding 9·5) may also inhibit zooplankton. The evidence, however, suggests that zooplankton and chironomids are important quality feeds in ponds but that they are grazed heavily, and thus their standing stock is low (Hepher and Pruginin, 1981). Thus, a comparison of ponds in Israel showed that although manure increased zooplankton and chironomid production (Table 1.1) this may not be as apparent in ponds stocked with fish, as any increase in density was matched by an increase in grazing. According to Hepher and Pruginin (1981) any increase in zooplankton and chironomids is dependent on an increase in the production of bacteria developing on the organic matter of the manure. This was demonstrated by Schroeder and Hepher (1979) who reported that zooplankton (*Daphnia* sp.) ingested small manure particles, digested the bacteria and then excreted the residual particles. It has also been reported that zooplankton can utilise the soluble fractions of manure (Hepher and Pruginin, 1981).

Table 1.1. Standing crops of zooplankton, chironomids, and bacteria in manured and nonmanured ponds with and without fish. (After Schroeder, 1974.)

	Without fish		With fish	
	Manured	Nonmanured	Manured	Nonmanured
Zooplankton (mg dry matter/m^3)	3·3–42·4	<0·055	0·34–1·3	<0·055
Chironomids (100's/m^2)	79–215	1–7	1–4	1–2
Bacteria (1000's/ml)	17–27	0·7–4·3	1·6–6·7	

1.5 The role of fish in integrated systems

1.5.1 General observations

The use of fish species feeding low in the food web, such as carps and tilapias, as opposed to carnivorous species, is a general feature of waste-fed culture systems. This allows increased production using only the natural foods produced in the ponds, and/or low grade supplementary feeds. Fish usually eat selectively in response to the range of available feeds, although they may be more flexible in their feeding when farmed. Polyculture, or mixed species culture, takes advantage of this diversity of feeding preference and is often important in allowing higher stocking densities and, hence, yields.

Polyculture of different fish species within the three dimensional pond environment may be compared with multiple cropping of land crops. The polyculture concept may be extended to include an integrated farm as a whole, with animals, plants, and fish all cultured together to optimise output and nutrient utilisation. This may be especially important where land is limited, and this approach may be the only practical means for sustainable small-scale agriculture in many parts of the developing world.

1.5.2 Practical considerations, monoculture/polyculture

Fundamentally, competition between fish species within a polyculture can be avoided if their feeding strategies are not the same. Maximum advantage might therefore be expected in polycultures of the most specialised feeders. For instance, Chinese carp polycultures seem to be more productive than those of the less specialised feeding Indian major carps.

A polyculture using a large number of fish species would be expected to best utilise the diverse environment of the waste-fed pond. In a Chinese carp polyculture, at least four kinds of carp are stocked together in pond systems; a surface feeder, grass carp (*Ctenopharyngodon idella*) that feeds mainly on higher aquatic plants, two pelagic mid water feeders, one of which, the silver carp (*H. molitrix*) prefers phytoplankton for food, and the other; bighead carp (*A. nobilis*), zooplankton.

A further group of bottom feeding omnivores which can be stocked include the common carp (*Cyprinus carpio*), the mud carp (*Cirrhina molitorella*) and black carp (*Mylopharyngodon piceus*). Whereas traditionally in China at least four or five species were used, it is increasingly common for more to be added in the hope of further increases of yield (FAO, 1983). Elsewhere, however, polycultures often contain fewer fish species than those used in China, where the practice began.

Many polycultures may use only two or three species but still achieve considerable improvements in yield over single species culture. The stocking of common carp, for instance, with one other species, such as silver carp or tilapia has significantly increased yields in Israel (Yashouv, 1971).

The expansion of polyculture in other places may be hampered by several factors. It is crucial to the success of polyculture to stock compatible fish of diverse feeding habits. This requires experience and an understanding of the ecology of local fishes if these are to be used. The use of local fishes may also be problematic if there is a lack of wild fry and/or hatchery technology. For these reasons the wholesale import of tried and trusted forms of polyculture has been attempted, often using the basic Chinese and Indian carps. By contrast, the simplicity of monoculture and the stocking of a single fish species, often of higher individual value, may often be very attractive.

In principle, one species cannot efficiently use all the available food in a waste-fed pond, though stocking omnivorous and less specialised feeding fish is better than using those more selective in their diet. However some of the highest yields observed in integrated systems have been achieved in monoculture. Yields of the Nile tilapia, *Oreochromis niloticus*, for instance, in waste-fed ponds, suggest such systems are practical. This species is particularly tolerant of poor water quality and thus allows the use of higher waste loads than are possible in polycultures that include more sensitive fish.

Other factors may practically limit polyculture. Hepher and Pruginin (1981) have identified that extra handling at harvest for polyculture compared with monoculture is expensive in terms of manpower and equipment. The different species of fish in a polyculture may also reach marketable size at different times, further complicating harvest procedure. In China, a significant factor in the high production achieved is the ready availability of labour for the complex and labour intensive polyculture strategies employed.

The use of polyculture may therefore be restricted by management considerations. The market value and acceptability of component species within the polyculture may also be a problem. A good example of this is the silver carp which, although commonly included in polycultures because of its efficient harvesting and use of phytoplankton, generally commands a poor price compared to other, less biologically useful, species.

1.5.3 Developments in polyculture

In spite of these possible drawbacks, progress and development in polyculture systems continue. Recently, more accurate assessments of the nutritional and living requirements of individual species have indicated more clearly the specific advantages of polyculture over monoculture and have identified the best ratios of species for use. The introduction of *Macrobrachium rosenbergii,* the giant freshwater prawn, into Israel, and consequent research into appropriate culture systems, is an excellent example. Originally considered a candidate for intensive pond monoculture (Mires, 1983), recent research suggests that the best overall economic return (though not necessarily the highest production) can be obtained in a polyculture with such finfish as tilapia and common carp. Contributing factors in the improved performance of stocks within the polyculture include reduced competition and cannibalism of the prawns (aided by stocking fewer, larger (1g) juveniles) and improved environmental conditions, associated with the feeding activities of the fish. Cohen and Ra'anan (1983) report that the prawn yields and average weights are independent of hybrid tilapia stocking rates in polycultures. Evidence from studies using different feeding/manuring strategies suggests that *Macrobrachium* prawns within polycultures feed mainly on the natural productivity of the pond which can normally be improved with low cost feed/manuring strategies.

Such findings are important where market demand is changing in favour of prawns, and their production is set to expand. Also, in a polyculture where fish production will provide for local protein consumption (tilapias, carps), *Macrobrachium* will supply external revenue; the use of such low cost inputs yielding a good return with very little commercial risk. The practical management difficulties of a high value prawn/low value tilapia polyculture can be minimised by culturing the fish in cages in the prawn pond. The low value tilapia are thus prevented from eating the costly prawn feed, whilst maintaining water quality by consuming excess phytoplankton. In addition, problems of excessive reproduction of tilapia and harvesting difficulties are minimised.

1.5.4 Polyculture—rationale for use in waste-fed systems

In general, polyculture will be advantageous in integrated waste-fed ponds for several reasons—

(a) Increased production and profit

The increased yield from fish ponds that can result from polyculture is subject to economic constraints. Shang (1981) states that polyculture will be viable regardless of yield if stocking a secondary high value species increases the total profit for a given pond area (e.g. stocking prawns in milkfish (*Chanos chanos*) ponds). Polyculture in this case may also reduce the risk of loss inherent in the monoculture of high value, but biologically sensitive, species. However, monoculture of a high value species will only be replaced by polyculture (possibly many less valuable species) if the change does not affect the total profit of the continued production. In the latter case, as the relative yields may not be predictable, the system may not be changed.

Box 1.4. Polyculture production; competition and yield.

An index of competition can be defined for species in polyculture as follows
(e.g.) tilapia yield in monoculture = 1500 kg/ha/yr
in polyculture = 1200 kg/ha/yr

$$\text{Index of competition} = \frac{1500-1200}{1500} = 20\%$$

Effects on profits of production

	Monoculture		Polyculture	
	Yield kg	Profit $	Yield kg	Profit $
Macrobrachium	1500	6000	500	2000
Tilapia	3000	3000	2000	2000
Carp	3000	2400	2000	1600

In this example, profits can be increased by changing either a tilapia or carp monoculture to a tilapia/carp, or tilapia/carp/*Macrobrachium* polyculture. If starting from a *Macrobrachium* monoculture however, polyculture would decrease profits.

N.B. This is a hypothetical example only!

(b) Improvement in pond environment

Mixed species culture may aid in the maintenance or improvement of water quality within the pond, allowing a higher stocking rate of fish. The blooming or super-abundance of the phytoplankton community in intensively manured systems can, for instance, cause water quality problems. The stocking of phytoplankton feeders, such as silver carp or some tilapia species e.g. the Nile tilapia, can help in their control. Similarly, the use of macrophyte feeders to control emergent weed growth is well documented. The presence of bottom feeding omnivores e.g. common carp may prevent anaerobic conditions occurring and can accelerate the resuspension and use of nutrients in the water column. Overstocking of these species may result in turbidity however, with detrimental results on primary production.

(c) Flexibility in integrated systems

Polyculture is particularly relevant when integrated into well established agricultural practices. Thus, the ratio of fish species can be manipulated to best utilise the particular waste available. This is illustrated by the pre-eminent use of grass carp (*Ctenopharyngodon idella*) in the Pearl River region of China where fish culture is more extensive and reliant on fodder crops for inputs (Edwards 1982[a]). The Chinese consider the grass carp to be a 'living manuring machine' since its high fibre intake and poorly digested diet result in large amounts of faecal waste being produced in the pond and further entering into the food web to feed other fish.

In contrast, in the region of the lower Yangstse River basin, a greater variety of by-products and manures are available and a more even balance of Chinese carps is stocked to use the waste more efficiently.

(d) Control of fry production

Wild spawning within the pond can seriously depress yields. Tilapia species are notorious for the 'stunting' of stocked fish as their off-spring compete with them for food and space within the pond. Stocking of predatory fish, such as the snakehead, *Channa striata,* in small numbers may control excessive reproduction and hence maintain the pond population in balance. In practice this balance may be difficult to achieve. However, the potential for fry control exists in many species. Even the polyculture of the bottom feeding omnivorous common carp with tilapia is reported (Yashouv, 1971) to control the fry production of the latter, although the method is subject to some conjecture.

1.6 Waste loading and environmental quality

1.6.1 Oxygen

The maintenance of adequate levels of dissolved oxygen is an essential part of achieving high fish production. Anoxia, or oxygen deficiency, has been identified as one of the major causes of fish kills in waste-fed ponds. If the oxygen content of the water in such ponds is analysed throughout a 24 hour cycle, a large variation will be observed. This is basically because of a change in oxygen production and use (respiration) within the pond. Phytoplankton, by way of photosynthesis during daylight hours, are the main producers of oxygen within the pond. Respiration by the plankton, heterotrophic and fish communities, is the primary consumer of oxygen throughout the cycle. Fertile ponds exhibit large fluctuations in oxygen content, usually reaching a maximum in the late afternoon after daylight photosynthetic production, and a minimum level in the early morning.

Clearly, food and oxygen production within the pond are linked. The correct balance of waste inputs and living communities, phytoplankton or bacteria, is thus important in preventing anoxia, whilst at the same time allowing efficient food production. Over addition of waste inputs for instance, encouraging a super-abundance of phytoplankton and bacteria, can cause a reduction in oxygen production over the daily cycle, and an increase in its consumption compared to less fertile systems. The former occurs because of autoshading which reduces light intensity to algae deeper in the pond and hence overall oxygen production. Thus, the importance of fish grazing on phytoplankton becomes apparent. Recent experimentation using the Nile tilapia has revealed more clearly the interrelationship between phytoplankton and oxygen production in waste-loaded ponds. Colman and Edwards (1985) found that the net production of the blue-green algae *Microcystis* was actually stimulated by the grazing activity of the fish; since the photosynthetic oxygen production in the pond is proportional to the net production of algae—both feed and oxygen reach a maximum value at the same time. An abundance of dissolved oxygen also stimulates food production by bacteria via the heterotrophic pathways. Conversely,, phytoplankton growth can be limited by the availability of nutrients, which are being made available by the activity of the heterotrophs. Thus, it is clear that both major feeding pathways are dynamically linked and essential to the maintenance of water quality.

Climatic conditions are also important factors in preventing anoxia. The prevailing temperature, for instance, affects the dissolved oxygen content. The amount of oxygen in water decreases as temperature rises; at high temperatures, not only is less oxygen available, but the fishes' metabolism increases and requires more oxygen. Sunlight, its duration and intensity are also important. Cloudy conditions, for instance, reduce light intensity and the lowered photosynthetic oxygen production can cause problems. Conversely, wind causes turbulence of the surface waters and allows greater diffusion of atmospheric oxygen into the water. Acute anoxia can often be traced to a combination of biological and climatic factors.

Fish species vary in their tolerance to conditions of low dissolved oxygen. Some cultured species are air breathers and these fish are least affected e.g. the snakehead (*Channa*) and catfish (*Clarias*) species. *Oreochromis* species (tilapia) depend on dissolved oxygen in the water but are less sensitive than the carp family (Cyprinidae) to conditions of low oxygen. Short term tolerance of low oxygen conditions is not the only factor of importance and should be carefully related to the culture situation. Sub-lethal effects such as reduced growth rate at low oxygen levels, even in resistant species, are potential problems which do not have the dramatic manifestation of the fish kill, but can be just as economically debilitating.

1.6.2 Toxic products

In intensive fish culture systems, where fish are stocked at high densities and fed large amounts of expensive feed, other aspects of water quality can also become problematic. In waste-fed ponds, fish are less densely stocked and potentially toxic waste products are usually absorbed into the natural food production before acute toxicity levels are reached. Ammonia, for example, which is an excretory product of the fish, and extremely toxic, is usually quickly taken up and utilised by the phytoplankton and bacteria in the water column. The blue-green algae in particular, by way of some unique physiological processes, are capable of assimilating large amounts of ammonia. Adequate water quality can be maintained even in intensive fish culture systems (Rijn *et al.*, 1984) which heavily load nitrogen into the system. A sudden die-off of the algal bloom or increase in fish density, which removes the algae by grazing, can also cause dangerous increases in ammonia. Hopkins and Cruz (1982), however, found depressed growth and survival of common carp in ponds that were heavily fertilised with chicken manure, and suspected this to be due to toxicity from ammonia. The well observed phenomenon of end of season inhibition of fish growth, in both intensive monoculture and integrated systems alike, may be correlated with the accumulation of toxic compounds e.g. sulphide in the sediments. The possibility of growth inhibition from chemicals produced by the fish themselves (pheromones) has also been suggested.

CHAPTER 2
MANAGEMENT OF WASTE-FED PONDS

2.1 General considerations

The management goal for a waste-fed pond depends on the primary objectives. In most cases the main goal is the maximisation of revenue, which often, but not always, coincides with the achievement of maximum production in such systems. Strictly speaking, production would be increased to the point at which further production would reduce revenue (e.g. if it became necessary to use increasingly expensive inputs), though in practice inputs are often limited by availability, or the farmer becomes less prepared to accept the risk of loss in higher cost systems.

In other cases, a primary objective may be to utilise wastes. Under these circumstances the addition of wastes would be continued to a level at which risks of overload and killing fish increased unacceptably, or to the point where wastes were not removed effectively.

The stimulation of natural food production in the pond, and its maximum utilisation by the fish, must be the primary consideration but this must be balanced by the maintenance of adequate water quality. The waste-fed eutrophic pond is more productive than a natural pond, but it is also less stable. Variations in water quality, especially dissolved oxygen levels, may be harmful to fish growth, or even cause acute mortalities. This 'instability' can be reduced by judicious management, however. The adjustment of inputs to the pond's biological and physical capacity to use them is part of this management. There is also evidence that a balanced variety of fish species may reduce the volatile and undesirable changes common among the pond food organisms which typically trigger large changes in water quality.

2.2 Addition of wastes

Traditionally, waste addition is managed by observing the colour of pond water and fish behaviour in the early morning. More inputs are added if the water is not green enough and hence providing insufficient food production. Amounts added are carefully controlled to avoid overproduction and anoxia; if fish are gasping at the surface of the water or if the colour is too intense, waste input is reduced and water may be exchanged. Zweig (1984[a]) observed that these methods are not easy to quantify and rely greatly on the skill and experience of the individual. Although used successfully in China, they may be difficult to transfer to other countries. Some operators recommend that if the arm is immersed in the water to the elbow and the palm of the hand can still be seen, there is a lack of natural food and more inputs are required.

The risk of anoxia through over-use of waste stands as the most important limitation to the amount of wastes that can be added to the pond. More quantifiable approaches can be made by using general loading guidelines, by measuring phytoplankton density directly and by determining the potential of individual wastes to break down and enter the pond's nutrient cycle and stimulate production. The matching of these more scientific methods with an appreciation of practical indicators can enable a rapid and effective management approach to be developed.

2.2.1 Biochemical oxygen demand (BOD)

The amount of oxygen required for the decomposition of a given waste in a specified time is termed its BOD. For the purposes of aquaculture, a 12 hour

Table 2.1. The 24 hour biochemical oxygen demand (BOD) for various inputs into pond culture of fish. (After Almazan and Boyd, 1978, Edwards, 1980b and Schroeder, 1980.)

	Material	% Dry matter	BOD g O_2/kg/24 hr at 30°C
	Pellets (25% protein)	90·0	140
GRAINS	Milled wheats/sorghum mixture	90·0	96
	Wheat grains	91·0	40
	Sorghum grains	88·0	18
	Chicken manure	95·0	20–40
MANURES	Field dry manure	36·0	10
	Liquid cowshed manure	12·5	7
	Liquid calf manure	9·0	5
	'Dry' human wastes	26·5	35–50
	Human sewage	2·0	2·5–3·0
FODDERS	Emergent aquatic weed	~8·0	5·4
	Floating aquatic weed	~8·0	6·3
	Submersed aquatic weed	~8·0	8·6
	Terrestrial fodder	~20·0	13·4

BOD ($BOD_{0.5}$) has often been found appropriate since decomposition has its most critical effect, decreasing dissolved oxygen in the pond, during the night. The value, which is waste specific, is highly correlated to the amount of dry matter (see Table 2.1). Decomposition is also temperature-dependent and thus, lower water temperatures will result in slower breakdown and hence lower values. As a rough guide, values reduce by 50% for every 5°C temperature drop. This facilitates the use of BOD values in estimating safe levels of organic matter that can be added to ponds, ensuring the minimum dissolved oxygen for fish is available.

Hepher and Pruginin (1981) suggest that safe quantities of manure added should be calculated with local conditions in mind, allowing for differences in type of manure, prevailing water temperature and the normal dissolved oxygen cycle. BOD values should be interpreted with care, however, since they take no account of nutrient composition and the rate at which the materials break down (see Fig. 2.1).

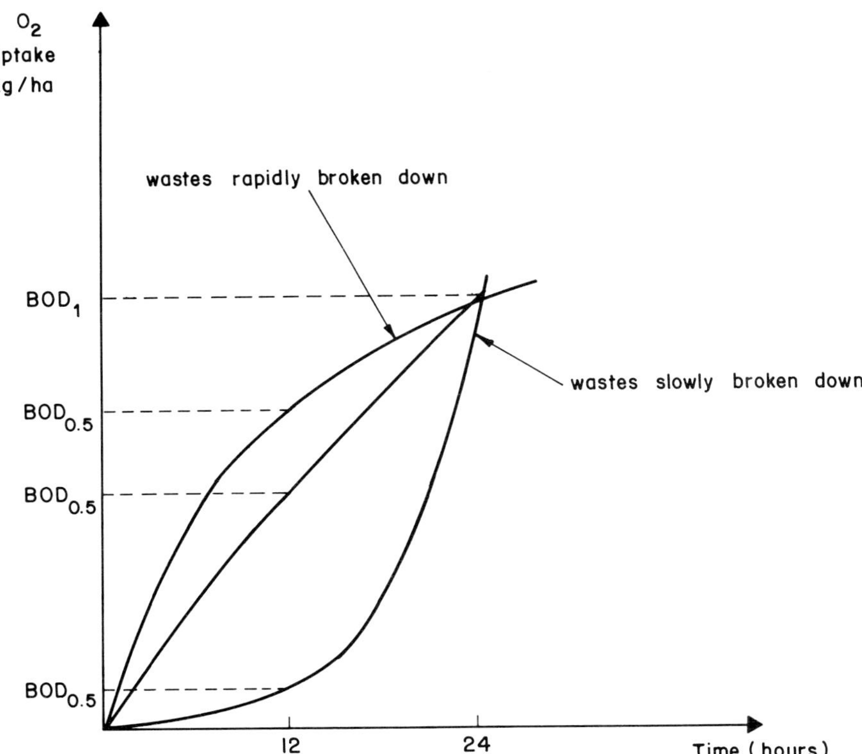

Fig. 2.1. Illustrating the potential pattern of breakdown of three wastes with highly different $BOD_{0.5}$ but the same BOD_1.

The action of these nutrients in stimulating primary production will also affect the oxygen regime. Indeed, Edwards (1982[b]) has estimated that the BOD of added manure is usually only a minor factor in overnight consumption of oxygen. The oxygen demand caused by the respiration of the phytoplankton developing as a result of the added nutrients appears to be more important.

Box 2.1. Use of BOD values in ponds.

A 1000 m^2 pond with average depth 1m holds 1000 m^3 of water.

If 100 kg of field dry manure is added to the pond, 24 hour BOD at 30°C is ~10 g/kg of manure/24 hr, then total oxygen requirement is 10 × 100 = 1000 g O_2/24 hr.

Thus, O_2 uptake is 1000g/1000 m^3 = 1 g/m^3 = 1 mg/l.

At 20°C, O_2 uptake will be ~50%, i.e. 0·5 mg/l. This is unlikely to load the pond excessively, unless the pond is already at the limit of night time O_2 depletion.

Thus, if the waste is added at 08:00, approximately 75% of the oxygen demand has built up by 02·00 the following morning i.e. 75% × 1 mg/l = 0·75 mg/l.

If the pond is normally at, say 2 mg/l at this time, dissolved O_2 would be reduced to 1·25 mg/l, which might cause mortalities. In practice, much will depend on the rate at which the waste breaks down, as part of this amount may be satisfied by photosynthesis during the day.

2.2.2 Chemical oxygen demand (COD)

Whereas the BOD value indicates the oxygen demand for respiration of micro-organisms acting on an organic substrate, the COD measures chemically the amount of oxygen required for the complete oxidation of a particular waste. However, the COD gives no information on the rate of oxidation of a particular waste and will not define, for example, the amount of oxygen taken up in 12 or 24 hours. If the wastes contain fractions that bacteria cannot readily oxidise, the COD value tends to overstate the effect of the wastes in the pond. A ratio of COD/BOD will however give useful information on the longer term oxygen requirements from waste added to ponds, since it is an indication of the rate of decay and thus oxygen demand over time. Thus if the COD greatly exceeds BOD, the waste will be oxidised slowly and can be expected to take up oxygen beyond 12 or 24 hours. If the values are approximately similar, the waste will have negligible long term effect.

2.2.3 Condition of wastes

In practical terms it may be difficult to add wastes at the rate which best fits the requirements for both natural food production and a healthy environment for the fish. Wastes, whether of plant or animal origin, are often of inconsistent quality or quantity throughout a growing season. Similarly, the requirements of the fish may change with time. If possible, integrated farming operations should be synchronised to meet the changing quantity of wastes and the changing needs of fish production. In terms of their value as material for heterotrophic food chains, wastes should ideally be finely fragmented and easily dispersible in water. This allows the maximum surface area for colonisation by heterotrophs and their rapid assimilation into the pond food web, and permits a more effective distribution through the water body. The value of very fibrous materials to heterotrophs compared to shorter chain carbohydrates has been questioned (Schroeder, 1978). Higher yields of bacterial cells were found on chicken manure than the more fibrous cow manure, for instance. Liquid manure is usually of a more homogenous and easily dispersed nature, which enables greater predictability of its action on the pond's food web. Age and method of storage will also affect conditions of wastes and their effectiveness as pond inputs.

Box 2.2. Use of dry matter (DM) loadings.

Recommended level, say 100 kg DM of wastes/ha/day.

A 1000 m^2 pond (e.g. 20 × 50m) can receive:

$$\frac{100 \times 1000}{10\,000} = 10 \text{ kg/day}$$

(a) Chicken manure; 95% dry matter (DM)

$$\text{weight required} = 10 \times \frac{100}{95} = 10{\cdot}5 \text{ kg/day}$$

(b) Cowshed manure; 12·5% dry matter (DM)

$$\text{weight required} = 10 \times \frac{100}{12{\cdot}5} = 80 \text{ kg/day}$$

This is a useful reference check in comparing productivities or fish yield resulting from different waste types at different loadings. Once the dry weight figure is obtained, the input of nitrogen (N) and phosphorus (P) can be calculated.

2.2.4 Frequency and timing of waste addition

The efficiency with which wastes are used in the food web appears to be affected by the frequency of their addition. Large occasional doses of wastes tend to stimulate a massive and unstable growth of bacteria, and phytoplankton, with the consequent risk of acute night time oxygen depletion. Growth of adequate grazers (mainly protozoa and zooplankton) to control these primary producers tends to lag behind. More frequent additions of manure allows the establishment of a more balanced food chain, ensuring that particular organisms do not reach excessive levels. Regular addition of wastes also minimises nutrient losses, since a larger, more balanced food web can absorb them more effectively.

When these matters are considered, it is not surprising that the pond systems that use the largest amount of wastes do so on a continual or very regular basis. Maximum loadings in such situations are estimated to be in the range of 100-200 kg of manure dry weight/ha/day (equivalent 70-140 kg of organic matter/ha/day) (Edwards, 1982[b]). These practices contrast strongly with traditional European practice in which the pond bottom is covered with manures, or wastes are piled in anaerobic heaps, practices which do not aid the oxygen regime of the pond or the distribution and use of nutrients within the waste. If applied daily, the timing of waste addition should be such that the capacity for the pond to use it effectively is at its greatest. Mid-morning, when oxygen content is rising fast through the photosynthesis of the algae, is probably the best time for manure addition.

The reasons for the notorious instability of waste-fed ponds are still poorly understood, but sudden changes in water quality are usually linked to the collapse of algal blooms. An acute reduction of dissolved oxygen and increase in ammonia concentration typically follow algal die off. Temporary nutrient deficiency is possibly a contributory factor in the collapse of algal blooms and this is best avoided by continual waste addition (see section 2.4). Eutrophic fish ponds tend to be dominated by blue-green algae, which have favourable characteristics in terms of feed quality and oxygen production. Colman and Edwards (1985) have found that heavy fertilisation, especially with organic wastes, coupled with seeding of *Microcystis* sp. will produce a stable and productive basis for tilapia culture. Nutrient analysis of waste-fed ponds can indicate the pathways of loss and conservation from the water. Edwards (1983) suggests that nitrogen is more likely to be limiting for bacterial and phytoplankton growth given the nitrogen to phosphorus ratio of livestock manures. Earlier work (e.g. Hickling, 1962), suggested that phosphorus was the main limiting nutrient, but this was on the basis of very extensive systems where biological fixation could provide the low levels of nitrogen required.

2.2.5 Waste addition and the weather

Prevailing conditions of temperature and light are important in determining the levels of wastes that can be safely added to fish ponds. Reduced ambient temperatures slow down bacterial decomposition of wastes, and the rate of growth of the food web generally. A build-up of organic matter in the sediments is possible if inputs are not adjusted, which can lead to undesirable conditions as and when temperatures do rise. Cloudy periods, resulting in reduced incident light energy and hence photosynthesis, also depress the pond system's ability to utilise large amounts of wastes. Seasonal adjustments in the amounts of wastes used are therefore necessary to prevent over or under-production of food.

2.3 Stocking and harvesting procedure

The stocking and harvesting procedures used to maintain the density of fish at an optimum level are typically labour intensive. In China, two main strategies are employed to ensure the most efficient use of ponds:

— rotation or mixed age stocking and harvesting, and
— 'multigrade' culture.

Rotation culture involves the stocking of several different size classes at one time, their regular harvest as they mature to market size, and their replacement with new fingerlings. A balanced sized community of fish is thus present all the time, ensuring that the diversity of the different feeding and living niches within the pond are used.

'Multigrade culture' is based on the knowledge that fish require increasingly more growing space as they grow. A polyculture of fish of the same size is thus gradually 'thinned down' and transplanted to other ponds as they grow. Zweig (1984[a]) recorded that up to five ponds in sequence are used, the last pond being stocked at low density and containing fish ready for marketing. This method is used generally in undrainable warm water ponds in South China, whereas rotation culture is practised in drainable ponds in temperate areas. Both procedures can be observed in other parts of the world although simple polyculture, where a single stock is introduced at the start of the season, is very common and requires considerably less management. A requirement for large sized fish is likely to encourage the multigrade method: where the market size is small, there will be less incentive for such intensive management. In general, the species yield within a well-balanced polyculture corresponds to the species stocked. Thus if 30% common carp is stocked, approximately 30% of the yield will be common carp.

2.4 Maintaining environmental conditions

The benefits of polyculture towards the management and improvement of oxygen regime and natural food production have already been stated. Water quality management is normally achieved by delivering nutrient inputs and by exchanging water. As nutrient loadings increase, the selective use of aeration becomes increasingly valuable, and may be quite cost-effective in permitting greater food production.

Typically, aeration is used to relieve the short term 'emergency' situation of early morning anoxia. Increasingly, strategies for a more regular use of mechanical aeration coupled with intensive feeding and production are being used. However, this approach in waste-fed systems, where stocking density and fish production are lower, will usually be uneconomic. The control of acute anoxia in ponds caused by sudden die-off of algal blooms is important in these systems. Superphosphate, added to the pond at a rate of 120-150 kg/ha, is reported to be effective in Israel (Hepher and Pruginin, 1981). The use of quicklime (200-300 kg/ha) followed by pig manure has also been found successful in preventing mass mortalities of fish in ponds during bloom die-off. It is important to use these remedies before the algae have died completely; they probably work by the calcium acting to precipitate the dying algae out of the water column and the phosphorus and pig manure respectively stimulating the production of new algae.

Box 2.3. Calculating stock rates and yields.

The previous discussions provide typical yield guidance, whether in kg/ha/yr or kg/ha/day terms. Thus, if using fertiliser and supplementary feed, in subtropical areas, the expected yield could be ~5000 kg/ha.

Stocking with fish capable of growing, say 250 g in six months in good conditions, this represents two total crops of 2500 kg annually.

Thus, if fish are harvested at 250 g (0·25 kg)

$$\frac{2500}{0\cdot 25} \longrightarrow 10\,000 \text{ fish/ha}$$

Allowing for *mortalities*, e.g. 10%

$$\text{stock} = 10\,000 \times \frac{100}{90} = 11\,100\text{/ha}$$

If *growth targets* are being defined, e.g. with the same fish, production rates ~30 kg/ha/day.

20 000/ha fish stocked at 50 g each ⟶ 1000 kg fish

If held, e.g. 100 days, additional weight ⟶ 3000 kg (30 kg/ha/day × 100 days)

$$\text{Final harvest weight} = \frac{3000 + 1000}{20\,000}(\text{kg}) \longrightarrow 200 \text{ g}$$

Waste-fed systems should be managed to optimise both the production of photosynthetic oxygen within the pond and its distribution throughout the water column. Such 'biological aeration' is most effective when the productivity of algae is at a maximum (Fig. 2.2). Colman and Edwards (1985) suggest the basic requirements for maintaining net oxygen production by the algae are:

— addition of sufficient nutrients to make the system light rather than nutrient limiting (many organic wastes are probably suitable though some may need supplementation with chemical fertilisers)
— sufficient grazing by fish to remove excess algae.

Excess algae cannot only reduce net oxygen production to below zero, but also make the system more sensitive to the destabilising effects of temporary nutrient shortage, and reduced light or temperature.

Typically much of the oxygen produced in eutrophic ponds during the day, when it may reach supersaturated levels, is lost by diffusion to the atmosphere, while deeper parts of the pond may be oxygen deficient. Circulation of water in the pond throughout the day can increase overall dissolved oxygen levels,

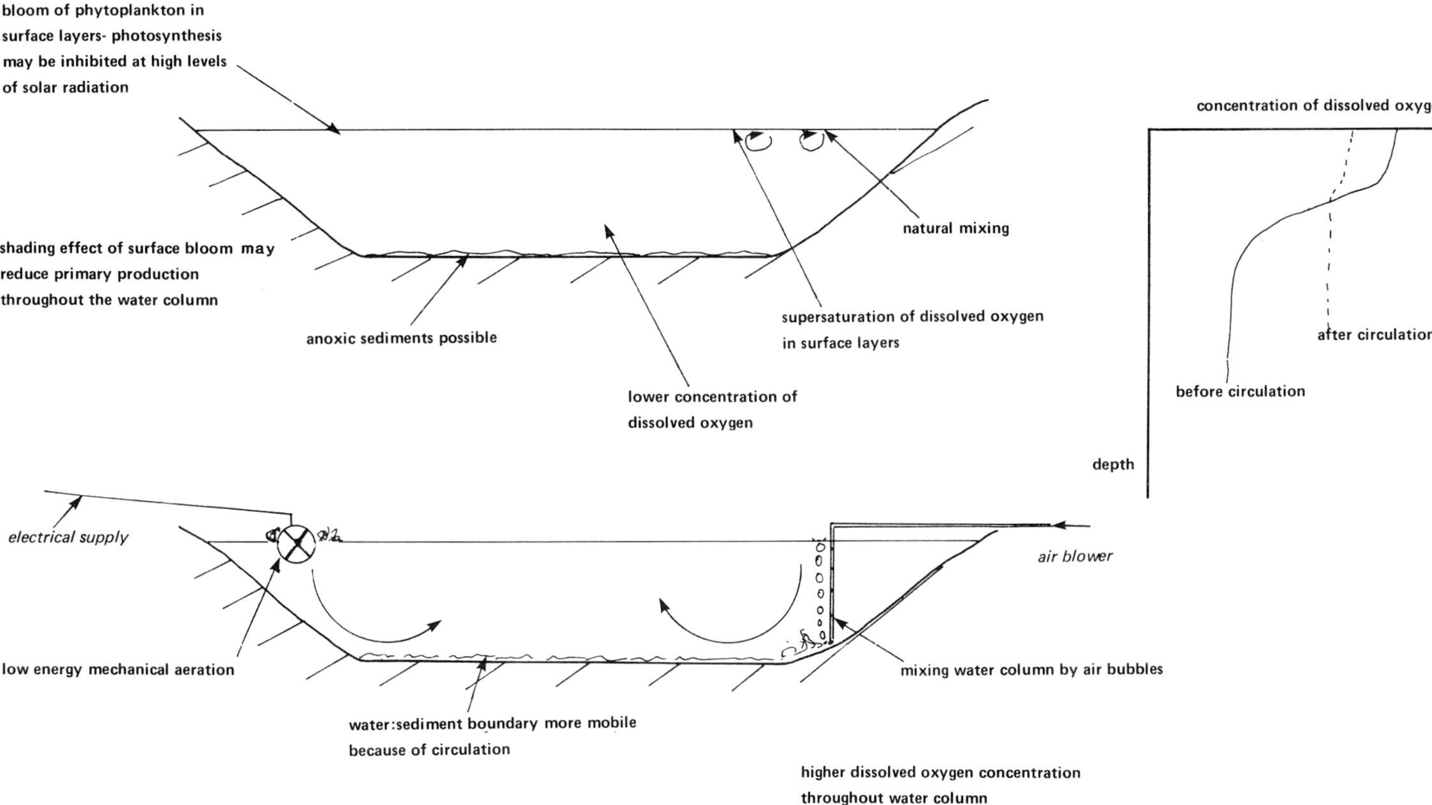

Fig. 2.3. Mechanical circulation for an improved dissolved oxygen regime in the waste-fed fish pond.

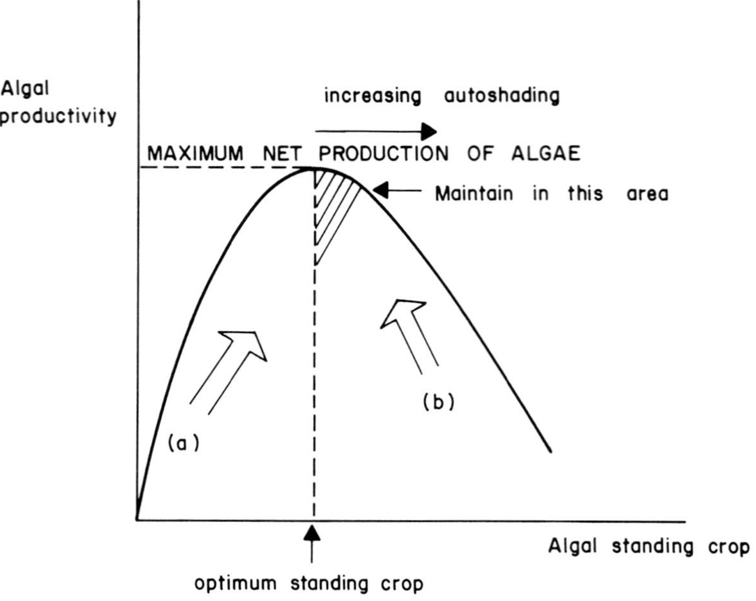

(a) increase algal standing crop

by – supplying more nutrients

and/or – harvesting some grazing fish

(b) decrease algal standing crop

by – increasing density of grazing fish

and/or – adding water to flush out excess

Fig. 2.2. Management of algal productivity for 'biological aeration'.

improving the oxygen regime throughout the 24 hour cycle. Simple mechanical and air bubble devices (Fig. 2.3) are a low energy method of achieving the circulation and may be relevant in the further improvement of yields in waste-fed ponds. The benefit to natural food production through physical aeration and water circulation, resuspending nutrients and aggregated detritus, may also be important, and supplement that from the movement and activity of the fish themselves.

CHAPTER 3
ANIMAL WASTES IN INTEGRATED FARMING

3.1 Introduction

When animal wastes are considered for use as fish pond inputs, several factors should be accounted for in the design of the system. Animal wastes may range from very concentrated effluents from intensive feedlot production to highly dispersed waste from extensive grazing, both of which may cause problems in their potential use for aquaculture. While organic loading should ideally be arranged to best fit the requirements of the pond system, fluctuations in quantity and quality of available wastes may create difficulties in management, and must be allowed for in attempting to integrate activities. Thus, fish ponds may receive more or less wastes than are optimal. In practice, constraints of land, water, labour and capital may also act to reduce benefits below those predicted from theoretical or experimental animal waste loading conditions.

The advantages of using animal wastes, however, may be considerable. When used as the only input in Israel, Schroeder (1978) reported that manures could achieve 75% of the yields attained by using supplementary feeding of grains and 60% of the yields possible with protein-rich pellets. Economic considerations arising from using animal wastes are discussed later (section 11.3); potential benefits are considerable, and may help to overcome some of the practical and physical difficulties of organising animal waste inputs to suit fish culture requirements. In addition, other advantages to their use, particularly with polycultures are:

— fish yields are found to be high and inputs relatively cheap. Animal protein is produced at a significantly lower cost than that of other farmed livestock (Wohlfarth and Schroeder, 1979)
— problems of manure disposal, often significant in intensive feedlot production, are avoided
— fish grown in manure-fed ponds tend to have a better texture and taste. Some researchers have found such fish to have a lower fat content (6%) compared to those fed on high-protein feeds (15%) or grains (20%) (Moav *et al.*, 1977).

3.2 Potential for integration

Livestock are an integral and important part of most agricultural communities. They are valued for their food and other products they provide. Despite the widespread use of agricultural tractors, nearly 85% of the total draught power used in agriculture throughout the world is still provided by animals (Smith, 1979). Traditionally, in many societies use is made of animal wastes for fertiliser and fuel.

Increased consumption of animal protein is an accepted indicator of improving affluence, and in areas of rising standards of living the demand for livestock products also increases (Brough, 1974). Intensive poultry and pig units, following Western practice, have been found to be relatively easy to establish in many developing countries. In general however, intensified production of livestock is not widespread in many parts of the developing world. Such methods tend to rely on large amounts of expensive concentrate and cereal based feeds. These may not be affordable on social or economic terms. However, integration with fish may provide an opportunity to improve the overall efficiency with which these inputs are used and it is often with such intensive forms of husbandry that fish culture has been most successfully integrated to date. The use by aquaculture of the large amounts of wastes generated by such systems is often found to be more reliable, acceptable and profitable than the alternatives of removal and use in terrestrial cropping, or disposal by dumping.

Integrated aquaculture has been considered particularly relevant to benefiting the rural poor. In more deprived subsistence economies however, the use of animal wastes in fish culture may compete with their use as traditional fuels and fertilisers. Also, since in these areas methods of livestock production are more extensive, opportunities for confinement of animals and the collection of their wastes may be reduced. Nonetheless, although optimal conditions for the use of wastes may not be possible or desirable, there may still be the potential for integration. Even draught and scavenging animals which have been found to compete minimally with man for food supplies (Smith, 1979), can be integrated with fish production. Simple improvements in management, i.e. night-time penning and supplementary feeding, may improve their productivity, allow greater control of wastes, and provide consequent benefits to integration.

3.3 Livestock production in the vicinity of fish culture

In integrated animal/fish farming systems, it is necessary for livestock to be confined for at least part of the day so that waste from feeding and excreta can be collected. For the following reasons, it may prove beneficial for livestock quarters to be either adjacent to or above the fish pond (Edwards, 1980[b]) (Fig. 3.1):

- waste products of the animals are likely to be used for fish production quickly. The nutritional value of the manure and fodder remnants are thus preserved because losses of nitrogen and energy due to fermentation and evaporation are eliminated
- feed residues can be eaten directly by the fish. Barasch *et al.* (1982) have estimated that such feed residues may reach 15% of that given to ducks

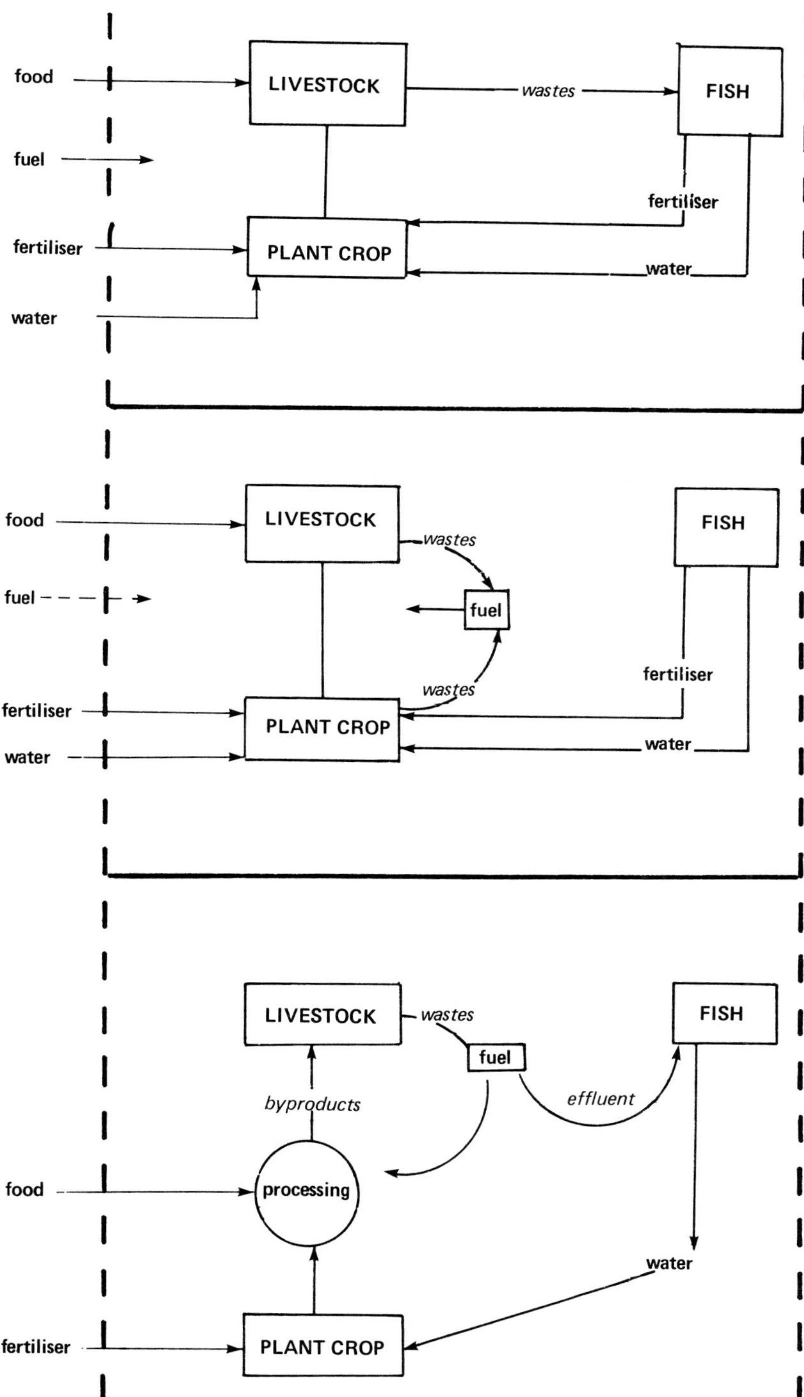

Fig. 3.1. Livestock production integrated with fish culture.

housed over fish ponds although use of pelleted feeds for the ducks reduces this 'loss' considerably

— costs of manure collecting, storing and transporting are obviated
— saving of land otherwise needed for housing the manure producing livestock (if the housing is above the fish pond)
— providing a good solution to problems of environmental pollution caused by animal waste
— improving the environment for manure producing livestock (Woynarovich, 1979). This appears to be particularly the case for ducks, though the improved returns on integrated farming may allow for the provision of better housing for other animals
— saving of livestock feed cost due to the natural food e.g. aquatic plants developing in the fish pond (in the case of ducks and geese). The operational efficiency of the farm is improved through better use of the manpower, joint use of feed storage, processing and transportation facilities (Woynarovich, 1979).

The number of animals kept per unit area of fish pond, as reported in the literature, varies from 500-10000 ducks per hectare, 1000-10000 chickens per hectare, and 15-200 pigs per hectare (Woynarovich, 1979; Nugent, 1978; Delmendo, 1980). Ducks and pigs are the farm animals most commonly raised in combination with fish (Wohlfarth and Schroeder, 1979).

According to Barasch *et al.* (1982) the reason for the wide range in stocking ratios observed for integrated animal/fish systems is related to the diversity of the different systems used but the principle factors are:

— climate, higher stocking rates have been reported from the tropics compared with temperate latitudes, and
— the fish species employed in the pond system.

For example, air breathing fish such as the catfish (*Clarias lazera*) are generally able to withstand poor water quality, and can therefore live in heavily manured ponds (Nugent, 1978). *Pangasius* spp. catfish and *Oreochromis niloticus* are reported to be raised in such ponds in Thailand (Wetcharagarum, 1980). In China, India and Israel however, carps, which are more sensitive to low dissolved oxygen concentrations, are more commonly stocked in manured ponds, and need a larger water surface to recycle a given quantity of animal waste (Delmendo, 1980).

3.4 Composition of animal wastes

Animal wastes and the nutrients they contain are valuable resources. If the proportion of input feed nutrients remaining in the wastes is considered, 72-79% of the nitrogen, 61-87% of the phosphorus, and 82-82% of the potassium, their potential value for fish culture can be appreciated (Edwards, 1980[b]). Waste output in the form of urine and faeces varies considerably in quantity and quality. Taiganides (1977) suggests species and total live weight of the animal to be the most important factors (Table 3.1).

Feed (amount and composition) and water intake, climate and management methods will also affect the value of waste to fish culture. In general it would be expected that animals fed high quality feed would have more nutrients in the waste. Similarly, livestock fed large quantities of feed would also have more nutrients since the faster passage through the gut would prevent efficient digestion and absorption (Edwards, 1982[b]). The quality and quantity of wastes from fattening units is also different from those from breeding animals.

The relative distribution of nutrients in solid and liquid wastes can also vary. Thus, higher levels of nitrogen (N) and potassium (K) have been found in urine (comprising 40% by weight of total waste excretion) than in faeces. A high phosphorus (P) content can be found in the faeces of animals, except pigs, which have considerable phosphorus in their urine (Delmendo, 1980).

Box 3.1. Calculation of theoretical nutrient input.

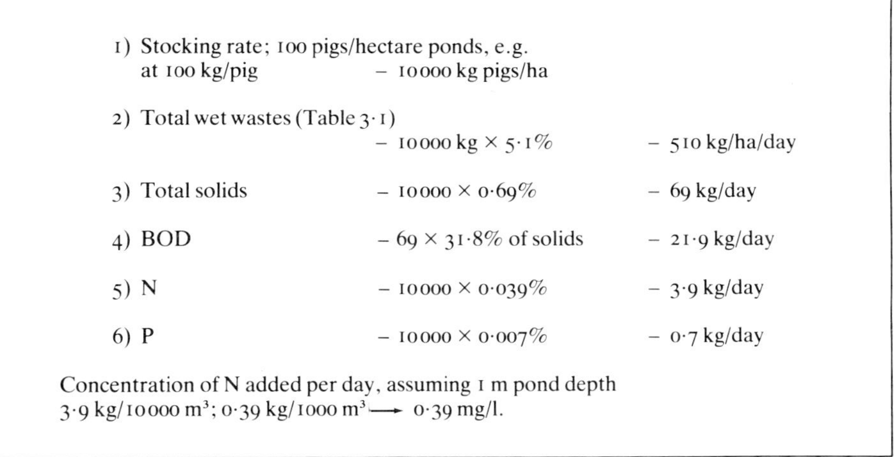

1) Stocking rate; 100 pigs/hectare ponds, e.g. at 100 kg/pig – 10000 kg pigs/ha

2) Total wet wastes (Table 3·1) – 10000 kg × 5·1% – 510 kg/ha/day

3) Total solids – 10000 × 0·69% – 69 kg/day

4) BOD – 69 × 31·8% of solids – 21·9 kg/day

5) N – 10000 × 0·039% – 3·9 kg/day

6) P – 10000 × 0·007% – 0·7 kg/day

Concentration of N added per day, assuming 1 m pond depth
3·9 kg/10000 m^3; 0·39 kg/1000 m^3 ⟶ 0·39 mg/l.

Table 3.1. Farm animal waste output. (Source: Meynell, 1982.)

	Pigs	Hens	Ducks	Cattle	Horses	Sheep
Animal weight (kg)	55	2	3	500	380	30
kg wet waste/animal/day	8	0·7	1·0	30	24	2·1
% faeces	45			70	70	66
% urine	53			30	30	34

Table 3.2. Composition of livestock wastes used in fish culture.

	As % of	Pork pig	Laying hens	Feedlot beef	Sheep	Cattle
Total wet wastes (TWW)	TLW/d	5·1	6·6	4·6	3·6	9·4
Total solids (TS)	TWW	13·5	25·3	17·2	29·7	9·3
	TLW/d	0·69	1·68	0·7	1·07	0·89
Total volatile solids (TVS)	TS	82·4	72·8	82·8	84·7	80·3
	TLW/d	0·57	1·22	0·65	0·91	0·72
Biochemical oxygen demand (BOD)	TS	31·8	21·4	16·2	8·8	20·4
	TVS	38·6	29·4	19·6	10·4	25·4
	TLW/d	0·22	0·36	0·13	0·09	0·18
COD:BOD		3·3	4·3	5·7	12·8	7·2
N	TS	5·6	5·9	7·8	4·0	4·0
	TLW/d	0·039	0·099	0·055	0·043	0·036
P	TS	1·1	2·0	0·5	0·6	0·5
	TLW/d	0·007	0·034	0·035	0·007	0·004
K	TS	1·2	1·7	1·5	2·4	1·4
	TLW/d	0·008	0·029	0·011	0·026	0·012

Note: TLW = Total live weight

Of the most commonly raised animals, wastes from poultry have the highest concentration of N, P and K followed by those of pigs and then cattle (Table 3.2). Fibre content in manure is also found to vary considerably. Cow manure was found to contain 60% total fibre (14-20% lignin, 30-40% crude fibre) compared to 45% for chicken manures (10% lignin, 25% crude fibre), and thus can be considered a richer source of carbon. Starch and reducing sugar content of both chicken and cow manure were both found to be less than 1% (Schroeder, 1980). Protein contained in animal manures is generally of a non-digestible nature and this will affect its value as a direct feed for fish. For instance, of the 25% crude protein in chicken manure, only 10% is pepsin digestible or true protein, available for digestion.

The quality of collected waste is affected particularly by three factors (Edwards, 1982[b]):

— whether or not the waste contains urine
— the amount of foreign materials e.g. bedding, or soil included
— whether water is used to clean pens and cool stock; the latter use may be considered under tropical conditions and may substantially dilute the

waste. Thus a reduction in dry matter content by a factor of 10 (from 10% to 1% dry matter) is common.

Bedding and uneaten food may considerably increase total dry solids and fibre content, whilst inclusion of soil in poultry manure may substantially increase ash content (up to 50% of dry solids). However, if the wastes are used directly in ponds, the same qualities of nutrients will be available; the addition of waste water or bedding merely acts to disperse the nutrients.

Box 3.2. What are typical nitrogen and phosphorus inputs into ponds?

Recommended inputs of nitrogen and phosphorus from both organic and inorganic sources vary widely; e.g.

		kg/ha/yr	
		Nitrogen	Phosphorus
from	100 pigs .	1423	256
or	30 kg ammonium sulphate/week	328	
	30 kg urea/week	718	
or	30 kg single superphosphate/week		109
	30 kg triple superphosphate/week		312

The output of these nutrients from the pond via harvested fish is quite small however.

e.g. assuming a 10000 kg/ha/yr yield of fish or approximately 1140 kg/DM/ha/yr

only 53·3 kg nitrogen
and 1·4 kg phosphorus are removed in the fish

Thus, the efficiency of nutrient transfer for a 10000 kg fish yield from the wastes of 100 pigs is =

$$\text{for nitrogen} = \frac{53{\cdot}26}{1423} = 3{\cdot}7\%$$

$$\text{phosphorus} = \frac{1{\cdot}425}{256} = 0{\cdot}56\%$$

3.5 Effects of storage

Taiganides (1977) has reported large losses of nitrogen (up to 90%) during storage under temperate conditions. In tropical climates, losses might be expected to be even greater.

Change and loss occur via chemical, biological and physical pathways, and are affected by collection and storage methods. Nitrogen is found in variable amounts as ammonia (NH_3), nitrate (NO_3) and nitrite (NO_2); loss by simple volatilisation of ammonia is accelerated by high temperature, wind and exposure to the environment. Bacterial denitrification, especially in anaerobic conditions, will also cause nitrogen losses. But the most rapid deterioration in waste quality, in terms of nitrogen and carbon loss, can be expected for wastes stored aerobically under tropical conditions at high temperature. Leaching of nutrients by rainfall and wind may also be causative factors of major loss.

Generally, the higher the temperature, the greater the surface area exposed to the atmosphere and the greater the storage time, the more substantial will be the loss of nutrients.

3.6 Constraints to animal and fish integration

(a) Animal production

The advantages of effective waste collection from intensive animal production have been suggested. Factors acting against such intensive production may therefore also constrain any development of integrated animal/fish systems. The high capital and operating cost of many intensive systems can preclude their use in developing countries, especially in the often disadvantaged rural sector. The absence of high performance stock, concentrate feeds and adequate veterinary and husbandry skills may also reduce the potential benefits of such systems.

(b) Competition of wastes

When the usefulness of animal wastes is well recognised, competition may exist for their use. Cattle and buffalo manure is widely used as a fuel in India, for instance, and chicken manure commands a high price in many places because of its value as a fertiliser. In such case, competing economic benefits must be considered carefully (see section 11.2), and the costs of reduced availability for other purposes must be included in proper assessment of integrating with aquaculture.

(c) Dislike of using animal wastes

A reluctance by farmers both to handle animal wastes prior to their use as a fish pond input and then to consume the cultured fish can be a sevère constraint on the successful promotion of integrated aquaculture. Direct addition of wastes by the housing of stock proximal to or over the pond may reduce the necessity to handle fresh, unpleasant wastes. The latter problem may be more difficult to resolve however; some racial and ethnic groups are more averse to the practice of eating waste-fed fish than others (see section 8.7). Some researchers note that once established, the practice can become more acceptable, both as the productive results become apparent and the good taste and marketability of the fish is proved. This problem can also be reduced, of course, if the production is sold outside the immediate area.

(d) The management problems of interdependence
When the demands of livestock management compete with those of fish production, practical integration of activities may be limited. For instance, the extensive production of ducks on fish ponds in Hungary has declined. This is partly because modern breeds of duck fed a high quality ration use the fish ponds less and benefit fish production less as a result, but is also because the need for concentrating management on the ducks limits the flexibility for managing the fish ponds. In other countries, land, water or labour availabilities may be more limiting than food or fish. As a result intensification of fish production may increase to the point where the physical integration with livestock becomes uneconomic. Thus the characteristics and limitations of a livestock waste-fed pond may not give a predictable and marketable enough fish harvest. The sensitivity of the natural food web to changes in climate and waste input may make such systems practically unmanageable if stocking density and fish yield are high.

(e) Public health constraints
The dangers of fish acting as vectors for human pathogens are still poorly defined, although certain studies are beginning to indicate the risks. Public health aspects of fish produced in human sewage are discussed later (section 8.7.1.) and might be expected to be more serious than those of animal/fish integration. However, Edwards *et al.* (1983[a]) have found that faecal coliforms and bacteriophage (*Escherichia coli*) levels were higher in pond water receiving duck waste than that receiving composted human waste; and that *Salmonella* was also present. Cloete *et al.* (1984) have made comparative studies of natural fish populations and those grown using cattle feed or effluent. Large numbers of bacteria, including potential pathogens were found on skin, gill and intestines but tissues and blood appeared sterile in both groups. This suggests that the consumption of such waste grown fish would not incur a health risk greater than for wild, supposedly 'clean' fish. In the light of recent evidence Hepher and Pruginin (1981) suggest that if the concentration of pathogens in the water is high they may be found in the fish tissues, rather than being confined to the digestive tract. Depuration, or the natural cleansing of contaminated fish in freshwater, and/or adequate processing should be adequate safeguards, although this may depend on local customs and practices. Velasquez (1980) has reviewed the diseases communicable or potentially communicable to man via fish and the water of animal/fish systems in the Philippines. Helminth infections are identified as being particularly dangerous, especially since the encysted form within the fish tissue can survive customary methods of preparation and preservation. In addition to a possible role in the spread of *Salmonella*, integrated fish systems may be implicated as the passive vectors of other bacterial diseases such as erysipelas and leptospirosis. Hubbert (1983) reported increased appearance of *Edwardsiella* spp. in the guts of carp fed digested cow manure, and noted that this type of bacteria has been associated with gastric infections in man, especially in the tropics.

(f) Waste quality

The variable qualities of animal wastes and their effects on fish production have been discussed. Certain types of wastes are more likely to adversely overload or underload the pond system, causing a reduction in fish yield.

Confinement methods for livestock often involve the use of bedding materials that alter considerably the available wastes for aquaculture. Nutrient concentrations, and the proportion of fibrous materials would typically change. Some materials, such as rice husk, which is often used for bedding broiler chickens, are known for their slow decomposition. Prolonged storage of such materials is thus required before their use as pond inputs is practical, which will also tend to reduce the waste's value.

In other circumstances, the mixing of easily biodegradable litter, and its rapid aerobic composting, may make greater quantities of quality input materials available.

The high nutrient quality of some wastes, may result in easy overloading of a pond system, with deleterious consequences for fish (see section 2.4).

CHAPTER 4
ANIMAL/FISH INTEGRATED SYSTEMS

4.1 Integrated pig/fish culture

4.1.1 Pig production

Both traditional and modern intensive pig rearing facilities have been successfully adapted for pig/fish culture. Some authors (Woynarovich, 1979; Delmendo, 1980) suggest that the pig rearing system should be designed to allow direct transfer of waste to the fish pond. The changing requirements of fish culture may however make some control mechanism for waste actually entering the ponds desirable. A simple form of temporary storage for waste can improve the flexibility of the system and may be particularly important in pig units producing large amounts of waste (Fig. 4.1).

A supply of water is required both for the pigs and for the regular cleaning of the sty itself. This may be considerable under tropical conditions (Devendra and Fuller, 1979). Nugent (1978) recommended a system of channels for loading the organic matter into the pond water, and stated that the enclosure itself should have a hard (concrete) base for maximum collection of food waste and faeces. The enclosure should be built to normal specifications for pig raising with respect to shelter, space, feeding and drinking facilities. In addition special consideration should be given to the needs of the breeding, nursing and fattening activities of pig farming. In tropical and sub-tropical conditions the siting of pig housing should be carefully chosen. Advantage may be gained from prevailing wind direction, both for cooling in very warm periods and protection during colder conditions.

Alternatively the sty may be constructed over the fish pond. Typically, the structure is supported on wooden stilts over the pond and provided with a lattice type of floor which permits the excreta and uneaten food to fall directly into the pond water without requiring the regular washing out and sluicing of the traditional concrete based housing. This type of construction may also have some cost advantages although this is largely dependent on the cost and availability of wood compared to concrete.

The pig sty can be built on the banks of the fish pond, and provided with a floor sloping towards the pond. Generally, housing built directly over the fish pond is used for pig fattening rather than for breeding. Experience at one of the largest pig/fish units in Asia, Farm Kirikarn, Central Thailand, suggests that pig pens and fish ponds are better separated. When housed directly over the ponds, the humid conditions are said to encourage respiratory problems and

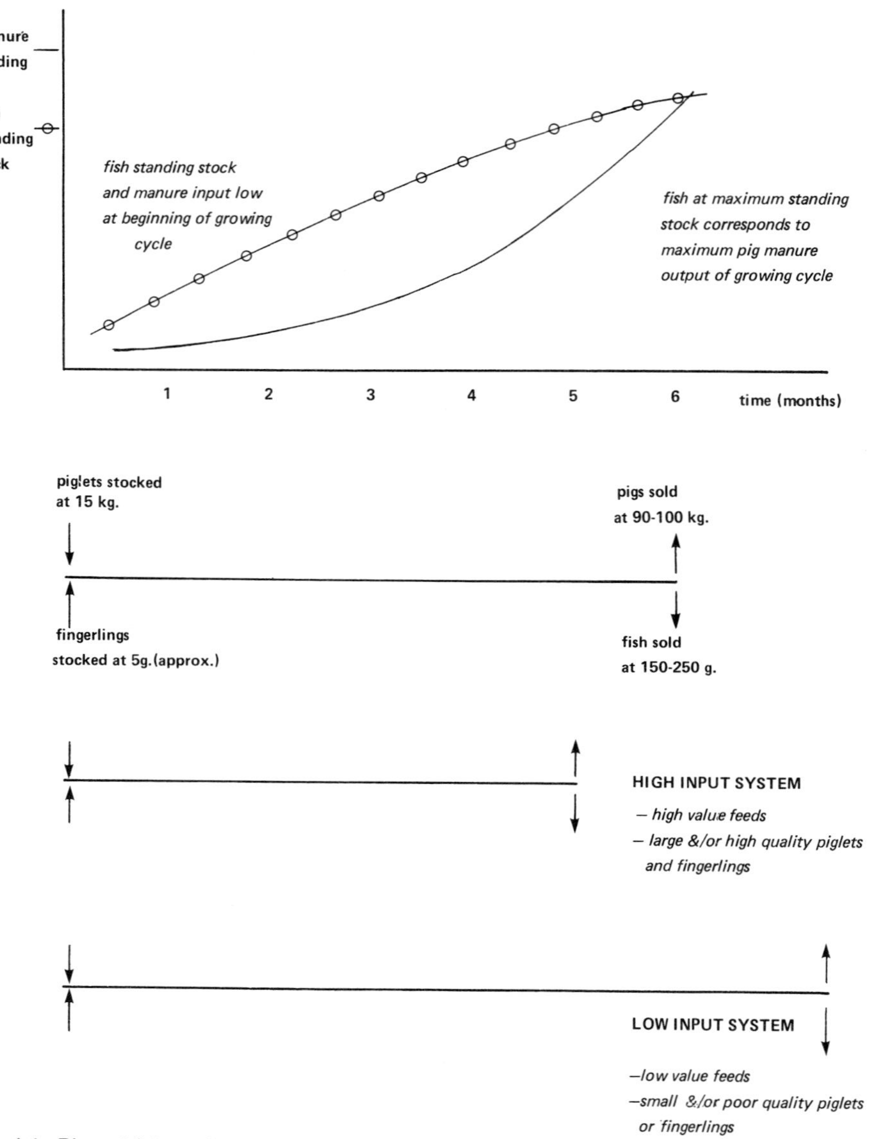

Fig. 4.1. Pig and fish production cycles.

farrowing difficulties. In addition, housing over the ponds can limit the use of chemicals for cleaning and medical purposes (Anon, 1985[a]).

Collection of waste for distribution to more than one pond is also possible. A simple system of concrete vault and channels can direct wastes to different ponds. Some systems use large quantities of pig waste without direct transfer to ponds. Large piggery units on some Hungarian state farms are often physically separated from the fish production ponds. This necessitates pumping the waste from slurry lagoons and transporting it by tankers to the ponds. Alternatively,

Woynarovich (1979) described three other methods used for the distribution of manures into fish ponds:

— soft manure is shovelled into a basket of parallel iron rods (~ 2.5 cm apart), suspended to a depth of ~ 10-20 cm from an outboard motor and washed out, diluted and dispersed by the turbulent current created by the movement of the boat
— using a pump built into the bottom of a boat the manure is shovelled into a hopper, diluted with pond water and then sprayed out into the pond through a hose equipped with an end plate so that the mixture is fanned out
— the dilution of the manure takes place on land and the distribution is carried out by hand from the shore or from a small boat. This method is used in small ponds where the use of mechanical equipment is not practical.

The methods aim to maximise the distribution of the manure over the pond surface and throughout the water column. The first two methods are only justified in terms of relatively large ponds.

Breeding stock

An important factor in pig/fish culture is the breed of livestock used. In India, landrace piglets are used (Jhingran and Sharma, 1980), while a large white-landrace cross is predominantly used in the Philippines (Cruz and Shehadeh, 1980). In Africa local breeds of pigs are adapted to pasture and not commonly kept in enclosures. European breeds such as Yorkshire and large white which are adapted to enclosures are more sensitive to tropical climates and diseases (Nugent, 1978).

Integrating pig and fish production

The procedure for pig/fish rearing can follow a number of patterns. In its simplest form it is the fattening of piglets (1·5-3 months old) to a marketable size (~ 60-100 kg) over a feeding period of between 5-6 months. This permits two production cycles per year, each corresponding to one fish culture cycle. Hopkins and Cruz (1982) reported that in the Philippines 10-15 kg weaner pigs were raised for 180 days according to normal practice. Feeding occurred at a rate of approximately 3·5-7% body weight per day, the pigs being then marketed at a mean weight of approximately 80-105 kg. Within one growing cycle of pigs, two production cycles (2 × 90 days) of tilapia (*Oreochromis niloticus*) can be achieved. More normally, one fish production cycle would compliment that of one pig production cycle over six months. As the pigs grow, their waste and food requirements increase as does the requirement for food of the growing fish; thus the balance in the pond can be maintained. High input systems or those using larger weaner pigs and fingerlings may have a reduced production cycle time. Similarly, the production cycle will be longer if pig or fingerlings are small or feed inputs are lower (see Fig. 4.1). Pig and fish production can therefore be complimentary.

Box 4.1. Sizing a pig/fish system.

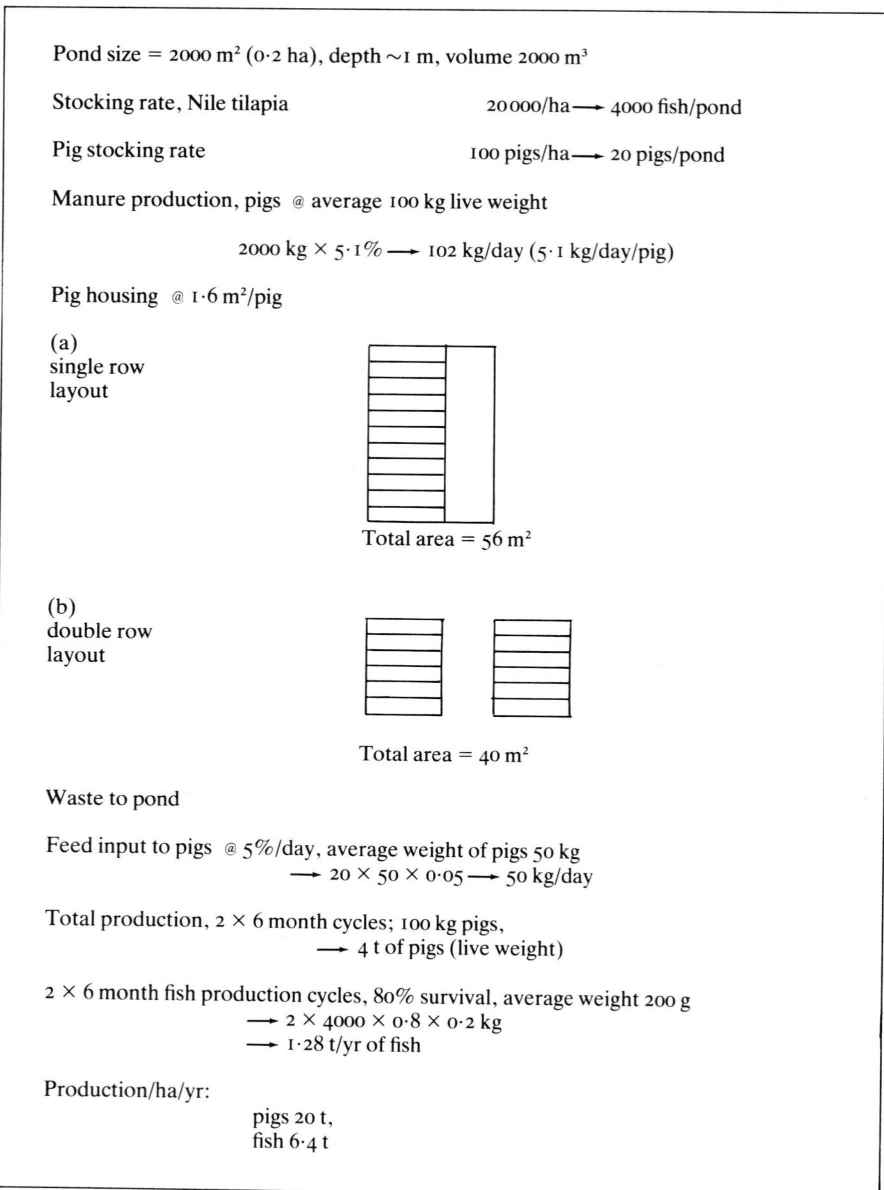

Pond size = 2000 m^2 (0·2 ha), depth ~1 m, volume 2000 m^3

Stocking rate, Nile tilapia 20 000/ha → 4000 fish/pond

Pig stocking rate 100 pigs/ha → 20 pigs/pond

Manure production, pigs @ average 100 kg live weight

2000 kg × 5·1% → 102 kg/day (5·1 kg/day/pig)

Pig housing @ 1·6 m^2/pig

(a) single row layout

Total area = 56 m^2

(b) double row layout

Total area = 40 m^2

Waste to pond

Feed input to pigs @ 5%/day, average weight of pigs 50 kg
→ 20 × 50 × 0·05 → 50 kg/day

Total production, 2 × 6 month cycles; 100 kg pigs,
→ 4 t of pigs (live weight)

2 × 6 month fish production cycles, 80% survival, average weight 200 g
→ 2 × 4000 × 0·8 × 0·2 kg
→ 1·28 t/yr of fish

Production/ha/yr:

pigs 20 t,
fish 6·4 t

Traditionally, in some areas, pigs are fed with uneaten or waste food from households or restaurants rather than balanced, concentrate feeds. Water lettuce (*Pistia stratiotes*) has been grown in ponds and after harvest mixed with rice bran. Water hyacinth (*Eichhornia crassipes*) has been chopped and cooked with abattoir wastes. The amount of green feeds that can be given should not be overestimated however; there is a trade off between the amount of grain used and the speed of fattening and, probably, the quality of manure for fish culture. In China, where grains have been fed sparingly to fattening pigs, between 0·5-1 kg grain/day/pig is used, together with 5-6 kg green feeds/day/pig (Plucknett and Beemer, 1981). Peanut cake, corn meal and soybean meal are also used when available at a low price (Le Mare, 1952; Delmendo, 1980). In Thailand, where the silver striped catfish (*Pangasius sutchi*) is commonly raised on pig manure, waste restaurant food is often fed to both pigs and fish. The feed formulations and daily feed requirements for pig rearing in the tropics are shown in Appendices 1 and 2.

4.1.2 Pig wastes and their use as pond inputs

The number of pigs used in systems should be related to the manure loading that the fish pond can support. The number of animals kept averages 60 pigs/ha of fish pond, although Edwards (1985[a]) reported up to 300 pigs/ha in systems raising the silver striped catfish. Woynarovich and Kunhold (1979) reported that one fattening pig produces on average 6·5-7·5 kg manure/day. Hopkins and Cruz (1982) reported an average fresh manure output per pig of approximately 134 kg during the first 90 day period of the rearing cycle and 307 kg for the second 90 day period. Tapiador *et al.* (1977) reported that in China 30-45 pigs are raised to supply adequate manure for each hectare of fish pond, but in addition many other inputs are also given. Also, as mentioned previously, the quantity and composition of organic manure varies according to feeds fed, age and total live weight, amongst other factors. A large variation has been found in the chemical composition of commercial piggery wastes both in Hungary (Woynarovich, 1979) and Australia (Hilliard *et al.*, 1979). This variation illustrates the effect of management on the quality of wastes.

4.1.3 Fish stocking densities and species composition

The composition of stocked fish and the initial stocking density varies considerably from system to system and depends on several factors, including the desired size of the fish at harvest; the initial weight of the stocked fish; the length of the growing season; the average water temperature; the fish carrying capacity and quality of the pond; and the local practices utilised (Nugent, 1978; Schroeder, 1980).

In the Philippines stocking of the Nile tilapia at 10000, 20000 and 30000 fish/ha, along with much smaller numbers of bottom feeding common carp and predatory snakeheads (*Channa striata*) has been attempted (Hopkins and Cruz, 1982). The higher stocking densities were used to investigate whether in-pond fry production of the tilapia could be reduced (since adults are territorial and require space to breed) and if a market size (60g +) could be attained. Delmendo (1980) reported that tilapia (*Oreochromis* spp.) are the preferred species for stocking in integrated pig/fish systems throughout Asia, and are used at rates of 25000-30000 fish/ha.

4.1.4 Yields

Fish production in pig/fish farming operations ranges from approximately 2000-5000 kg/ha/6 months. However, the silver striped catfish (*P. sutchi*), an air breather, can yield 15 t/ha when grown in monoculture.

Hopkins and Cruz (1982) found that the highest yields of marketable fish in a tilapia, common carp and snakehead polyculture were achieved using a high stocking density (30000/ha) and a culture period of 6 months (4-5000 kg/ha/6 months). Higher total yields (6000 kg/ha/6 months) were gained when fewer tilapia were stocked (20000/ha) without predators, although a high proportion of the production was of small individuals and unmarketable.

In the USA, fish production in pig/fish systems is reported to range between 17-22 kg fish/day (3060-3960 kg/ha/6 months), in a polyculture largely comprised of Chinese carps. The wastes from between 39-66 pigs/ha of pond were used (Buck *et al.*, 1979).

4.2 Integrated poultry/fish culture

4.2.1 General consideration

Small-scale poultry production is a typical feature of rural communities the world over. In many places this practice has been supplemented or replaced by more intensive methods. Both intensive and extensive poultry production have been integrated successfully with fish culture; the low purchase and feeding cost per individual bird make poultry a common investment by poor farmers. Poultry are usually readily available and their productivity can often be improved with simple and cheap management. Some supplementary feeding and immunisation against disease can reduce losses dramatically, and partial confinement can improve manure availability to fish culture. The value of poultry in the form of eggs, meat and feathers, gives them a flexibility advantageous to subsistence and market orientated farmers alike. Most types of poultry give manures, in the form of a combined urine and faeces output, of high value. Not only is the nutrient quality high (see Appendix 3) but it is also low in fibre and of a finely fragmented nature. Moreover, if advantage can be taken of the natural integration of many forms of fowl and water, e.g. ducks and geese, maximum dispersion of wastes throughout the fish pond can be achieved.

As with pigs, poultry can be intensified in the tropics successfully without resort to large import bills for concentrate feedstuffs. Payne (1981) suggests that sorghum and millet grains or various types of oilseed extraction by-products can be used in the dry tropics, whilst in wetter areas, processed cassava, sweet potato, yam, breadfruit and sago can be substituted for cereal grains. Replacement of protein concentrates can be made with grain, legume and tree crop by-products.

There follows a description of various poultry/fish systems, but similar methods can be employed for any form of domesticated fowl, e.g. pigeons, turkey or quail.

4.2.2 Duck/fish culture systems

Background

Combined fish/duck culture has been widely practised in the Far East and Eastern Europe, especially in Hungary (Woynarovich, 1979). Higher fish yields as well as improved feed conversion and reduced fat content are said to have resulted through integration with ducks. The ducks, on the other hand, were reported to be more healthy and cleaner, and the quantity and quality of their feathers also improved.

The rationale for raising ducks on fish ponds is stronger than many other animal/fish systems, since the pond can provide living and foraging area for the ducks as well as the fish. Indeed, ducks may be considered an extension of the polyculture within the pond further using space and food sources unavailable to the fish. According to Hepher and Pruginin (1981) two methods of raising ducks have been developed:

(a) extensive raising, in which only small amounts of supplementary duck feed is provided and the number of ducks is limited by the food they can find in the pond water. The amount of manure contributed to the pond and its effect on fish yield are also limited. This method is usually employed in Europe, where approximately 150-500 ducks are held per hectare of pond surface (Woynarovich, 1979)

(b) intensive raising, in which the ducks are fed at the same rates as on land and held at a much higher density per unit of pond area. Therefore, higher amounts of manure and uneaten duck feed (estimated to be 10%) are loaded into the fish pond, and consequently higher yields can be obtained. This method is usually employed in Africa, where duck stocking densities are approximately 1000-2500/ha (Wohlfarth and Schroeder, 1979) and in the Far East where the stocking densities range between 750-13 125 ducks/ha (average 4640/ha) (Sin, 1980).

Edwards (1985[b]) has further defined duck/fish systems to include a semi-intensive mode of duck production; this differentiates between the methods on the basis of fertilisation. Extensive and intensive methods of duck production rely completely on natural feeds and complete feeds respectively, with no reliance on fertilisation. The semi-intensive mode is characterised by fertilisation to produce natural feed e.g. ducks grazing in fertilised rice fields or fish ponds in addition to wild sources of natural feeds.

The seasonality of fish culture in Europe would be expected to limit the amount of waste addition and hence intensive stocking of ducks over fish ponds. Tropical conditions allow higher loadings, and thus advantage can be taken by using more intensive methods. A premium on the availability of land in the densely populated parts of Asia has also stimulated the development of more intensive methods of raising ducks.

Duck rearing facilities

Ducks can be fed and sheltered either on floating rafts or on stilt structures built over the pond surface. In both cases the faecal material and uneaten food falls directly into the pond (Nugent, 1978). Chen and Li (1980) reported that in Taiwan the duck house is built alongside the fish pond using either a bamboo frame with straw roofing, or, cement poles, wooden beams and asbestos cement corrugated roofing. The cement or brick floor is covered with wooden gratings so that the ducks are isolated from their excreta and/or uneaten food. (see Fig. 3.1.)

Jhingran and Sharma (1980) reported that in India, the ducks were allowed to range freely over the pond water by day and sheltered at night in a floating duck house made of bamboo matting over empty oil drums, positioned close to the pond bank, where the young ducklings and breeding birds were usually kept (Nugent, 1978). Sin (1980) described another system used in Hong Kong, where a small section of the pond is enclosed by wire netting and connected to duck sheds of between 18-75 m^2, built either on adjacent land or in the pond. The sheds are usually walled with wire netting floors; the enclosures are provided with wooden platforms as duck resting places. He noted that each shed of 20 m^2 was capable of accommodating approximately 400 ducks (i.e. 20 ducks/m^2). In Thailand, duck houses have been constructed over the pond using only bamboo; a fence of bamboo strips enclosing approximately half of the pond area confined the ducks over the water, but allowed fish access to the entire pond. Edwards (1983) recommended five ducks per m^2 of floor space with this type of housing. Nylon netting has also been used for fencing but this was found to be less durable and effective. Regular maintenance of duck fences and houses is required. Ducks should be restricted to a relatively small area of pond if they are being raised in the intensive mode, to avoid the expenditure of too much energy in swimming (Edwards, 1985[b]).

The design of duck housing should ideally allow total input of duck wastes into the pond. Confinement of the ducks to the pond, without access to the bank sides ensures this but, especially on a larger scale, may make for difficulties in effective management. Ducks need access to shelter, whether from wind, rain or sunlight, and in intensive systems large amounts of feed have to be made available to them. If ducks are allowed access to the bank sides for feeding and/or shelter, a high proportion of the feed and droppings are wasted and the structure of the pond itself may deteriorate rapidly (Barasch *et al.*, 1982). Reinforcing of bank sides has been attempted in Hungary using various artificial fibres but this may not be generally cost effective.

Duck raising

The rearing of ducks through their life cycle to market size is frequently divided into specialised activities:

- production of one day old ducklings
- nursing of young ducklings
- rearing of advanced ducklings up to market size.

Details of the specialist requirements of producing day old ducklings and rearing young ducklings are given in Appendices 4 and 5 respectively; this section discusses the rearing of advanced ducklings to market size, the main activity of interest in integrating farming.

The Hungarian method of production carefully and gradually conditions high quality ducklings, which have been hatched under artificial conditions, to water in small ponds before release into larger production ponds. Ducks have been observed to eat small fish and so they are not released into nursery ponds (Woynarovich, 1980).

Breeding stock

Ducks may be raised specifically for meat or egg production; markets may be the deciding factor as to which is suitable. Edwards (1983) considered egg laying Khaki-Campbell type ducks as being most suitable for small-scale subsistence production in Thailand on the basis that:

— the ducks will furnish a daily supply of good quality protein, i.e. eggs
— surplus eggs are easier to sell than whole ducks, and
— ducklings need to be replaced only every 18 months, as opposed to three months for fast maturing meat ducks.

In addition an even manure input is maintained over a longer period which is beneficial for fish production.

Different duck strains are used according to traditional practices and availability. In Taiwan the ducks raised are either the egg producing native duck or the meat producing mule duck (a cross of the native mallard *Anas boscas* and the drake of the muscovy *Cairina moschata* (Chen and Li, 1980). In Thailand the Khaki-Campbell (*Anas platyrhychos*) is raised (Lawson, 1981), whereas a Khaki-Campbell-Bengal runner cross is favoured in India (Jhingran and Sharma, 1980). Sin (1980) reported that in Hong Kong either the local introduced Thai breed and/or a Taiwanese hybrid (male Peking-female Denmark cross) is raised. In the Philippines and in Central Europe the Peking duck and the various hybrids of this strain are the most commonly cultured.

Barasch *et al.* (1982) note that in general the genetic quality of the ducks integrated with fish culture is often inferior. The development of improved quality ducks may therefore be important for the viability of the integrated system as a whole.

Edwards *et al.* (1983[b]) have also observed reduced laying rate and poorer economic viability when locally available ducks as compared to introduced strains are used in integrated small-scale systems in Thailand. Commercially, most duck eggs are produced in more extensive systems in Thailand, where the ducks can derive a portion of their food requirements from the environment and thus reduce feed costs. The total confinement of ducks for manure collection for fish culture when the profitability of egg production is marginal may not be feasible in such cases. Poor laying rates were also found to occur during the rainy season and in very hot weather. In addition, moulting every 2-4 months lowers egg production as does disturbance in the form of dogs barking, car lights etc.

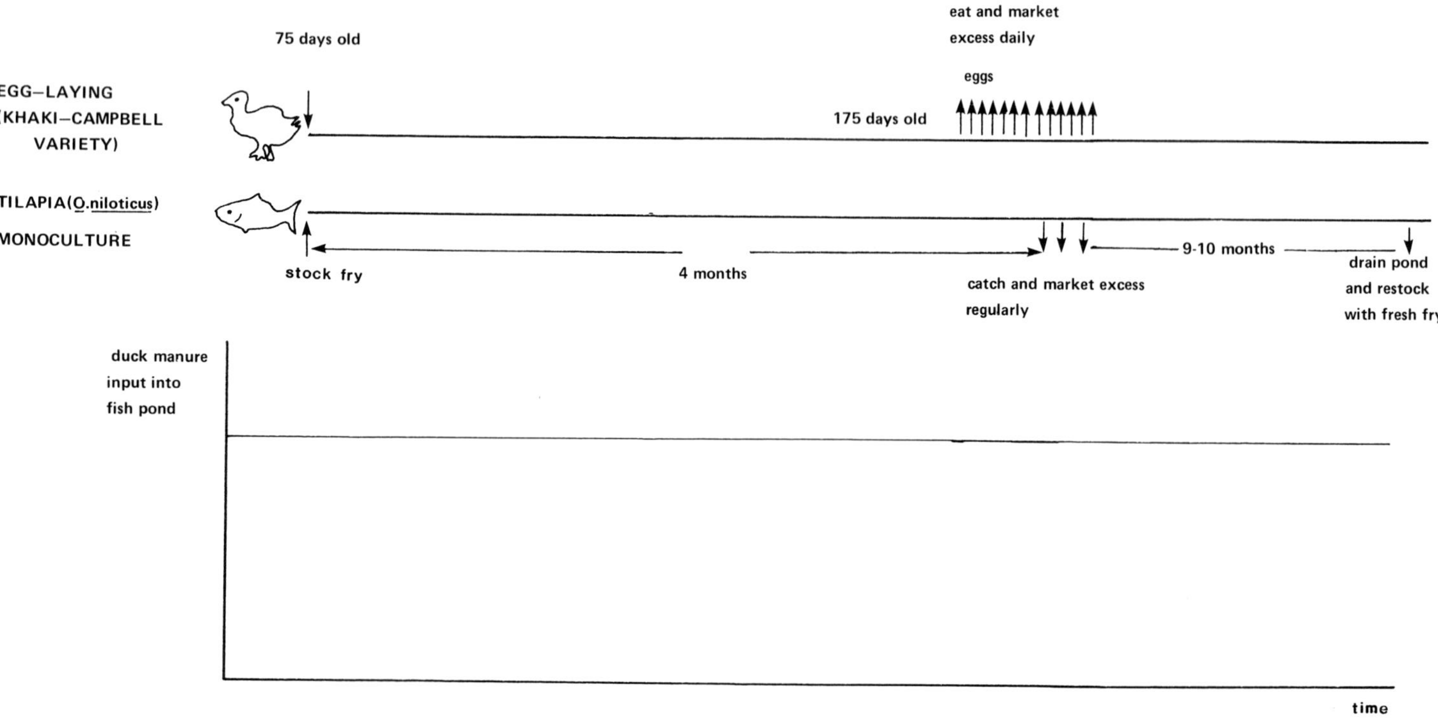

Fig. 4.2. Production cycle of egg laying ducks integrated with tilapia.

Some of the recently improved production strains of ducks are less inclined to range over the pond. Such ducks, produced in Europe and Asia, are normally kept in confined spaces or water runs so that efficient utilisation of expensive artificial feed is ensured. The present meat production strains are reported to reach a marketable size of approximately 2·3-2·8 kg within 48-52 days. Traditional strains attain a carcass weight of around 2·0-2·4 kg within 55-56 days. Descriptions of typical production cycles are found in Figs. 4.2 and 4.3.

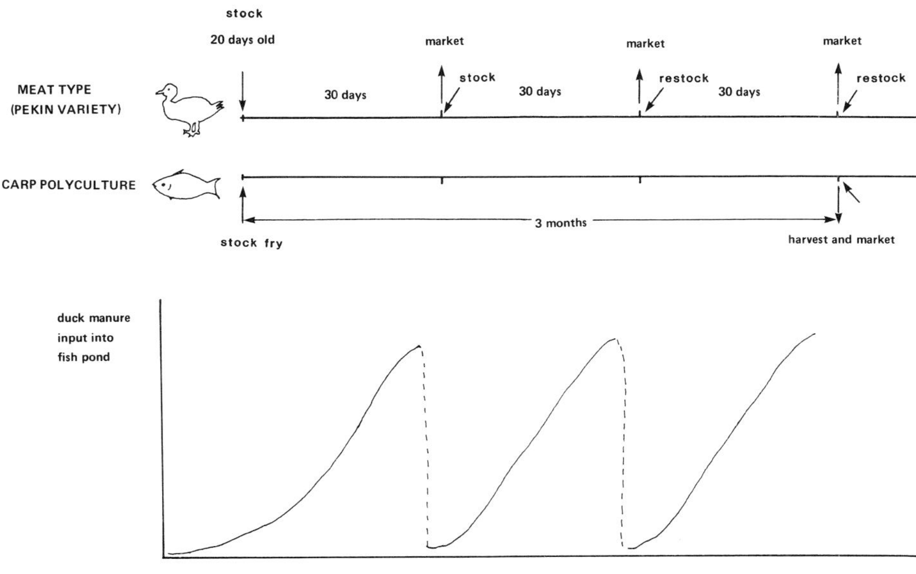

Fig. 4.3. Production cycle of meat type ducks integrated with carp.

Duck feeds

Commercial duck feeds, as used in Hungary, contain only 14-15% protein (N × 6·25) and any outstanding protein requirements are derived from the natural food in the ponds (Woynarovich, 1979).

Numerous authors have found that ducks will consume frogs, tadpoles, mosquito and dragon fly larvae and aquatic weeds which are generally not eaten by commonly stocked fish. Ducks can be fed with pellets, using demand feeders; the changing composition of feeds for different stages within the duck life cycle as used in Hungary is given in Appendix 6. Woynarovich (1980) has reported that to prevent pelleted feed digestibility from decreasing, the crude fibre content of the diet should be kept below 6-7%. He has also suggested that home made feed mixtures including chopped lucerne, clover, nettles or other suitable tender fresh plants mixed with suitable bran or crushed grain can reduce production costs under Hungarian conditions. However, duck growth on such feed is slower and its preparation is time consuming. Muscovy ducks may have advantages over other types of ducks in this instance, especially for small-scale systems, in that feed can include more fresh plant material grown on the farm. Edwards *et al.* (1983[b]) found that preparation of suitable feeds for small-scale integrated production of laying ducks can however be problematic. Diets containing a large proportion of paddy rice were found to be unsuitable for high yielding Khaki-Campbell type ducks which had access to little extra food from the ponds. A diet better balanced with respect to protein, energy and fibre was devised (Appendix 6). Feeding twice a day, (0·14 kg/duck/day) a stable egg laying rate of 80% was achieved under small farm conditions.

Duck manure—chemical composition and fertiliser potential
Duck manure had been reported to contain approximately 57% water and 26% organic matter (Table 3.2).

A single duck is reported to produce 7 kg of fresh manure over a 36 day period, of which approximately 4-6% is converted into fish flesh.

The effect of stocking ducks on fish ponds, or the regulated use of their waste, will vary widely with their age, size and production method. Fast maturing meat ducks (Fig. 4.3) will have an increasing manure output during their growing period; this should be managed to avoid over addition to the pond and consequent problems of fish production. The management of a mixed age flock of ducks can alleviate this problem. Conversely, egg laying ducks will supply a more constant input of waste into the pond since their feed input and manure output is more constant (Fig. 4.2).

Fish stocking density and species composition
Carp polycultures are commonly used to exploit the pond's natural food when enriched by duck manure. In Hungary a growing season of three summers produced fish of a marketable size of 1·0-1·5 kg. The polyculture stocked comprised approximately 1000 silver carp, 2-3000 bighead carp and 1500 common carp/ha, and received the wastes from approximately 300-500 ducks (Woynarovich, 1979). Assuming survival at 60%, the actual fish yield is thus approximately 900 kg/ha/yr, or 1·8-3·0 kg fish/duck/yr.

In Hong Kong the integrated duck/fish system is also based on stocking a mixture of carp species. In addition however, around 30% of the fish stock is comprised of grey mullet (*Mugil cephalus*) and up to 43% of tilapias of mixed species including hybrids (Table 4.1).

The number of ducks employed in this system varies between 750-13 125/ha of fish pond. The cold season followed by a warm spring can cause problems to fish/duck integrated systems in Hong Kong. Ducks are still produced in this period despite the pond's reduced capacity to absorb their waste. Fish kills are therefore common, especially if more than 3000 ducks/ha are stocked, as the warm weather begins. Chen and Li (1980) reported that a more complex polyculture system is practised in Taiwan in which around 10800 fish/ha are stocked together with 2200 mule ducks (Table 4.2). Lawson (1981) noted that in Thailand 200 m^2 ponds are stocked with 800 fish (Nile tilapia) fingerlings and 30 Khaki Campbell ducks each.

Table 4.1. Stocking density of different species in pure fish polyculture as compared with duck/fish and goose/fish integrated systems in Hong Kong. (Source: Sin, 1980.)

Fish species	Stocking density (no. of fish/ha)					
	Polyculture		Duck/fish		Goose/fish	
	Range	Average	Range	Average	Range	Average
Grey mullet	4800–18000	11900	1750–18750	7910	4630	4630
Silver carp	420–2570	1470	460–4500	1250	930–1290	1110
Bighead	600–2250	1420	300–3750	1590	860–1390	1130
Grass carp	625–1875	1600	500–3750	1760	2315–3220	2770
Common carp	1500–5450	2800	680–6920	3150	2315–2800	2560
Tilapias	130–10500	3740	1200–35290	12090		

Table 4.2. Stocking and harvesting of fish in a one hectare duck/fish farm raising 2200 mule ducks in Taiwan. (Source: Chen and Li, 1980.)

Species	Stocking			Harvesting		
	Size	Number	Time	Number	Weight (kg)	Time
Grass carp	15 cm	300	Feb-Mar	180	324	Oct-Dec
Bighead	13 cm	300	Feb-Mar	255	382	Oct-Dec
Silver carp	13 cm	1200	Feb-Mar	960	864	Oct-Dec
Mullet	2·5-4·0 cm	3000	Feb-Mar	1650	495	Oct-Dec
Sea perch	5 cm	200	Feb-Mar	160	144	July-Dec
Common carp	4 cm	1000	Feb-Mar	900	540	Oct-Dec
Eel	34/kg	40	Feb-Mar	32	5	Dec
Walking catfish (*Clarias batrachus*)	12/kg	72	Feb-Mar	56	17	Dec
Tilapia (mainly *O. niloticus* × *O. mossambicus* female hybrids)	600/kg	4740	Mar-Apr	10150	2900	June-Dec

Box 4.2. Sizing a duck/fish system.

Pond size = 2000 m² (0·2 ha), depth 1 m, volume 2000 m³

Two types of operation:

a) European carp production =
Carp stocking rate, say 500/ha → 100 carp/pond
Duck stocking rate 300/ha → 60 ducks/pond
Duck waste output, ducks @ averaging 2 kg weight,
(Waste rate ~0·096 kg/day per kg duck)
60 × 7 kg/36 days → 11·5 kg/day
Duck housing @ 10 ducks/m² – 6 m² (e.g. 4 m × 1·5 m)
i.e. 1·5 m out from pond, 4 m along bank
Conversion of waste to fish (4–6% quoted) say 5%
11·5 × 5% → 0·58 kg/day
Thus average carp weight gain → 0·58/100 → 5·8 g/day
Using a 200-day growing season, this could provide up to 1·16 kg growth per fish, or on an annual yield basis, 580 kg/ha/yr

b) S.E. Asian tilapia production =
Nile tilapia stocking rate, 20000/ha → 4000 fish/pond
Duck stocking rate, 8000/ha → 1600 ducks/pond
Duck waste output as above, assume mainly egg laying, waste estimated @ 60% of those in (a) 1600 × 0·096 × 60% → 92·2 kg/day
Duck housing @ 20 ducks/m² – 80 m², say 2 m × 40 m (say 2 units)

Egg production @ 60%	– 960 eggs/day
Feed input, 0·14 kg/duck/day	– 224 kg/day
Conversion of wastes to fish, 5% × 92·2 kg	→ 4·86 kg/day
Average daily weight gain of fish	→ 1·22 g/day
Using a 365 day growing season	→ 445 g/yr
Annual yield basis	→ 8·9 t/ha/yr

N.B. Tables 4.1 and 4.2 show how a polyculture would fit in this system.

Hopkins and Cruz (1982) found that a higher stocking density of the Nile tilapia (20000/ha rather than 10000/ha) produced a significant improvement in yields over a range of organic loadings. Barasch *et al.* (1982) similarly found higher yields with a more densely stocked polyculture of tilapia hybrids and carp, but that a reduced individual size of fish resulted.

Yields

Just as the pond system's capacity to absorb waste in tropical systems is high, so are the resultant fish yields. The highest average fish yields recorded on a daily basis are those of Barasch *et al.* (1982) at 36·5 kg/ha/day in Israel using around 1500 ducks/ha (extrapolated to 9 kg fish/duck/yr). Over a similar period Hopkins and Cruz (1980) and Cruz and Shehadeh (1980) reported 18·7 kg/ha/day in the Philippines using 750 ducks/ha (extrapolated to 9·1 kg fish/duck/yr).

Practically, yields on an annual basis tend to be lower. In Vietnam, raising ducks at 1000-2000/ha is said to improve average fish yields from 1-5 t/ha/yr compared with unfertilised systems (Delmendo, 1980), (~2-4 kg fish/duck/yr).

In Taiwan, fish yields of approximately 3500 kg/ha/yr without additional fertilisation or supplementary feeding have been achieved by raising up to 1500 ducks/ha (Chen and Li, 1980) (~2·3 kg/duck/yr). Using between 2500-3500 ducks/ha/yr, fish yields of between 2750-5640 kg/ha have been reported in China (Delmendo, 1980). Fish yields of over 10 kg/ha/day have been achieved in Africa with a duck/fish polyculture system of the Mossambique tilapia (*Oreochromis mossambicus*) and *Clarias lazera* (Nugent, 1978). Production of the Nile tilapia in small farm ponds (200 m^2) has yielded more than 7000 kg/ha/yr on an extrapolated basis (Edwards *et al.*, 1983[b]). The ponds were stocked at high density (4 fish/m^2) to produce high yields of small fish principally for home consumption. In temperate climates yields are lower. Probst (1934) observed yield increases from 236 kg/ha/yr to 417 kg/ha/yr when ducks were integrated with two age classes of common carp in Europe. When tench (*Tinca tinca*) were included in the polyculture further improvements were noted.

4.2.3 Goose/fish culture

Intensive fattening of geese over fish ponds has been practised in Hong Kong for over 25 years (Sin, 1980). The management techniques are similar to those of duck/fish (see above).

Geese raising facilities

According to Sin (1980) geese can be fattened in covered wooden sheds built over the pond water and provided with a wire-netting floor supported by a wood framework so that the goose droppings enter the pond directly. Each shed of approximately 360 m^2 is divided by a central corridor (1·5 m wide) with several compartments in which the geese are held. Metallic drinking troughs filled from suspended perforated water pipes are provided. The geese are allowed access to the pond via a sloped platform.

Breeding stock

Three types of goose, strains of the Chinese goose (*Cygnopis cygnoides*), are used for fattening: the flathead goose, yellow goose and the black goose.

Over a fattening period of around 17-20 days the geese increase in weight by more than 1 kg to reach market size of 3·0-4·5 kg.

Newly stocked geese are allowed to bathe for a few days in pond water before being transferred to fattening pens (7 m^2) each of which is able to hold between 50-70 geese of similar size.

The feeds compounded of sorghum, wheat and broken rice mixed in a ratio of 10:15:17 by weight are fed at a daily rate of about 40 kg per hundred geese. In addition, tender grass (12 kg/day) is fed to achieve a feed conversion ratio of approximately 7 or 8:1. The reported annual yield is approximately 2·25 t geese/ha.

Yields

For a production of 2·25 t (~500) of geese, yields of fish when raised as a polyculture of tilapias, common carp and grass carp, reach up to 3689 kg/ha (Table 4.1).

Potential

Geese have also been raised more extensively on fish ponds in Thailand. Compared to other poultry, inputs of fresh or concentrate feed can be reduced if adequate amounts of grazing are available. However, a greater tendency to remain on dry land will reduce their manure input into the pond. If geese are confined over the pond and fed forage by means of cut and carry, greater benefits to fish culture would result. Geese in small numbers on fish farms may also fulfill a security function, since large adults tend to be territorial, noisy and aggressive.

Box 4.3. Sizing a goose/fish system.

Using a 5000 m^2 pond (0·5 ha), with tilapia, common carp, grass carp polyculture

Stocking rate of fish, say	18000/ha tilapia	– 4000/pond
	2500/ha common carp	– 1250/pond
	1500/ha grass carp	– 750/pond

Goose stocking at 500 produced/ha/yr = 250 geese/pond/yr

Assume 6 crops of fattened geese/yr = ~40 geese stocked, which requires 1 pen of 7 m^2

Daily feed requirements:

Using sorghum/wheat/rice at 10:15:17 ratio

For 40 geese require 40 kg/100 geese –	3·8 kg sorghum/day
	5·7 kg wheat
	6·5 kg rice

Say 4 kg, 6 kg, 6·5 kg respectively

Grass input @ 12 kg/100 geese → 5 kg grass/day

Typical output sizes of fish, etc, assuming single annual crop

Tilapia 4000 × 80% survival × 300 g	– 960 kg
Carp 1250 × 85% survival × 500 g	– 530 kg
Grass carp 750 × 85% × 500 g	– 320 kg
Total	– 1810 kg
Annual fish production	– 3620 kg/ha

4.2.4 Chicken/fish culture

Background

Intensive production of broiler and egg laying chickens is now common in many parts of the world. The literature regarding chicken/fish integration generally relates to this form of 'battery' or deep litter production. The birds are typically fed complete diets in pelleted or mash form and the manure is used fresh or as dried poultry waste. The concentrated nutrient content of chicken manure favours its use as an agricultural fertiliser and animal feed supplement in many countries (Goddard, 1981), and its high value may therefore preclude its use in integrated aquaculture.

Chicken rearing facility

Most chicken housing is built on land rather than over the pond. There are broiler raising systems in South East Asia, however, that use stilted bamboo-type sheds directly over the pond. Typical battery cages may be used for egg layers and broilers, although the latter are more commonly raised on litter. A maximum density of 11 birds/m^2 is used for growing broilers in the Philippines (Hopkins and Cruz, 1982). In Indonesia the chickens are kept in battery houses erected on elevated land or the fish pond bank. Part of the chicken house protrudes over the pond raised with wooden poles, with split bamboo walls and supported by cement foundation (Djajadiredja *et al.*, 1980). In chicken/fish farming systems employed in Thailand, sheds of corrugated iron roofing material, with an earthen base are used for broilers (Wetcharagarum, 1980). Inevitably many of the advantages of integration are lost in such systems, as the manure may be left for long periods of time before use, with consequent deterioration in quality, and wastes are not easily transferred to ponds.

Chicken raising

Typically, young egg laying chickens are reared in nursing sheds for up to four months after which they are transferred to raising sheds. The egg laying period starts after the fifth month and continues for up to an average age of 22 months (achieving a monthly production of approximately 21 eggs per animal), when replacement by new stock takes place (Wetcharagarum, 1980).

Broiler chickens may attain a market size of 1·1-1·4 kg in about 49 days with average food conversion ratio (FCR) of around 2·6 (Hopkins and Cruz, 1982).

Chicken feed

Chickens integrated with fish culture are generally fed commercial rations, which have a balanced high protein content. Fed either *ad libitum* or twice daily, feeds are often 'diluted' by additional low cost feeds such as broken rice and maize (Wetcharagarum, 1980). In totally integrated systems, i.e. egg laying chickens housed over the pond, it has been estimated that 20% of the feed (Djajadiredja *et al.*, 1980) and almost all the excreta ($\sim$ 40 g/day) enter directly into the pond water.

Management of waste output

Ponds receiving chicken wastes will be subject to variation in their loading, as food intake and waste output increases over the life cycle of the birds. As with duck production, this variation will be greatest with meat or broiler bird production. Fluctuations in waste input may be reduced by using a production flock of mixed age. In the Philippines, Hopkins and Cruz (1982) used a flock of three size groups; as one group reached market weight after about seven weeks and was marketed, another group of chicks was bought to replace them. Manure output from mixed sized classes still varied from 11-31 kg dry matter (DM)/day for 1000 birds but this compared well with birds raised on a single age flock whose waste output varied between 1-45 kg DM/day.

Stocking ratio and fish yield
In a three month experimental period Hopkins and Cruz (1982) have illustrated an increasing fish production with numbers of chickens stocked of up to 5000 broilers/ha of fish pond. A polyculture of the Nile tilapia, common carp and snakeheads was used. Net fish yields were highly variable when only low quantities of chicken manure were added but reached an optimum of around 20 kg/ha/day for a loading of 100 kg DM/ha/day. Carp yields were highest at approximately half the chicken waste loading optimal for the Nile tilapia.

Burns and Stickney (1980) reported comparable yields in the USA of 16·2 kg/ha/day of *Oreochromis aureus* at 4000 chickens/ha of pond (1·5 kg fish/chicken/yr).

Box 4.4. Sizing a chicken/fish system.

Using a 2000 m^2 pond (0·2 ha)

Stocked with tilapia @ 16000/ha = 3200 tilapia/pond

Intensive chicken @ 5000 broilers/ha = 1000 chicken/pond

Waste output, averaging 2 kg dry matter/day/1000 birds
(Wet waste output = 2 × 100/44 @ 56% water)
– 2·0 kg DM
– 4·8 kg/day wet waste

Chicken housing @ 8/m^3 → 125 m^2, say 5 m wide × 25 m long

Chicken production @ 1·2 kg average weight/50 days – 8·4 t/yr
(1000 × 1·2 kg/50 days × 7 crops/yr)

Feed input @ FCR crop 2·6:1 – 21·8 t/yr, 60 kg/day

Uneaten food input to ponds, estimated 5% of chicken feed intake → 3 kg/day

Fish output @ ~ 10 kg/ha/day → 740 kg/yr
→ 290 g/fish average weight at 80% survival

Conversion of waste (DM) to fish (excluding feeds) = 0·99 = 1

4.3 Integrated ruminant/fish culture

4.3.1 General considerations

Ruminant/fish culture systems are rare. Most ruminants, animals having a complex stomach, are raised extensively, and so their wastes are less easy to manage and use. Their manures tend also to be bulkier and poorer in nutrients (Tables 3.1 and 3.2), factors that may limit their usefulness for aquaculture. The population of ruminants, however, is often high, even in otherwise poor countries. Despite low productivity by Western standards they often fulfill important roles in providing draught power, meat and milk while accepting poor or variable diets.

4.3.2 Prospects for confinement of ruminants

Partial or temporary confinement of ruminants may permit some beneficial integration with fish culture. More intensive practices for the management of ruminants are being introduced, and these are potential areas for more productive integration. A growing demand for milk in South East Asia for example, has stimulated the development of intensive dairying, and the consequent availability of concentrated wastes for fish culture.

The pressures to earn foreign exchange in developing countries have for many years stimulated cash cropping of primary fibre and foodstuffs. Large quantities of the low grade and fibrous by-products usually result, and can often be used by ruminants as a direct feed. Successful attempts have been made to include sugar by-products (De La Hunt, 1981) and green bananas (Ruiz, 1981) for instance as supplementary or basal ruminant feeds.

In the tropics, pasture availability may be seasonal, and certainly its value as a feed may vary considerably throughout the year. The use of locally available agricultural by-products may thus alleviate the consequent fluctuations in animal productivity. Indeed, this intensification may also reduce the environmental damage caused by ruminant overgrazing. Utilising crops residues, and even whole crops grown specifically for animal feeding, may thus become the basis for feeding stalled ruminants and so the potential for supplying aquaculture with wastes may improve.

In some areas of the developing world, ruminant production may be traditionally intensive and small-scale, as is the case with goats in Java and this may be particularly useful for integration with aquaculture.

4.3.3 Cattle/fish culture

Confinement

Modern intensive methods of cattle production for milk or beef usually involve the collection of wastes as a slurry. In Israel, the use of this slurry has been investigated as a fish pond input. Cattle wastes are allowed to collect in a pit under the slatted floor of the housing. The material contains about 10-12% dry matter (Hepher and Pruginin, 1981). The manure slurry is usually pumped out and often water is added to facilitate this, reducing manure to only 2-3% dry matter. Open corral methods of confinement inevitably result in higher initial losses from the wastes, especially when the floor is earthen and exposed. Improvements in the value of these wastes for aquaculture can be attained using relatively simple techniques (see section 9.3.2).

Stocking ratios and yields

Under controlled experimental conditions, workers in Israel (Moav *et al.*, 1977) have shown that high fish yields are obtainable using cattle slurry. Using a mixed age polyculture of carp and tilapias, yields of over 30 kg/ha/day were observed in a 126 day period. The cattle waste inputs into the pond were gradually increased over the period, from the equivalent of 57 dairy cows to nearly 400/ha of fish pond. Higher fish production was recorded when the fish were stocked more densely (32·7 kg/ha/day of fish when 18000 fish/ha were stocked compared to 31·5 kg/ha/day when only 9000 fish/ha were stocked). Overall, a feed conversion ratio of 2·7 was achieved, a figure comparable to that found when expensive grain was used. Hepher and Pruginin (1981) found a slightly poorer food conversion ratio (3·0-3·5). Schroeder (1974) reported fish yields in Israel of approximately 10950 kg/ha/yr within experimental ponds, receiving only cattle manure. Subsequently, Wohlfarth and Schroeder (1979) also in Israel, observed yields of around 14600 kg/ha/yr. These yields correspond with those found in Brazil (Carvalho *et al.*, 1979) where the wastes from calves (aged between 1-180 days) are used in fish culture. Yields of 3977 kg/ha were obtained over 130 days (11 166 kg/ha/yr) with a tilapia hybrid (male *O. hornorum* crossed with female *O. niloticus*).

Hepher (1975) has also correlated yields in manure ponds with the stocking density of the fish. A linear relationship was found up to 9300 fish/ha although these yields are only likely with optimum management.

Box 4.5. Conversion efficiency in cattle/fish systems.

Waste input into pond system from 60–400 animals/ha, say 250/ha;

dairy cattle slurry is ~79% water, 21·6 dry matter (DM)

If waste output is about 0·89% of total live weight/day (average live weight ~200 kg)
then slurry input is 250 × 200 × 0·89
= 445 kg/ha/day
or 93·5 kg DM/ha/day

If production of fish ~30 kg/ha/day
Conversion, DM slurry:fish = 93·5:30
= 3·1:1

4.3.4 Buffalo/fish culture

General considerations

The domesticated water buffalo, *Bubalus bubalus* is found throughout the tropical regions of Asia, the Mediterranean area, South America and the Caribbean (Ruskin and Shipley, 1976). It has a deserved reputation for production of meat and milk on poor quality, low nutrient vegetation. Buffaloes are kept mainly for draught purposes in subsistence farming systems throughout Asia; they are the traditional work animal in rice growing areas and are usually preferred to cattle for this purpose.

Physiologically, buffalo seem better adapted to the wet tropics than most breeds of cattle which are more suited to drier climates.

Observations of higher populations of gut bacteria and protozoa in buffaloes compared to zebu cattle (Chalmers, 1974), suggest a more efficient digestion of various feeds. Also, as faecal microbes may be an important direct source of protein for fish when manures are used as pond inputs, buffalo manure may therefore be a preferable input to that of cattle. On the basis of nitrogen content, however, buffalo manure contains approximately only half of that found in cattle waste (Edwards, 1983). This is no doubt a reflection of the different feeding and management regimes which are normally used. The dairy cow is likely to be much better fed than a buffalo under grazing conditions.

Recent research has shown that only poor yields of the Nile tilapia (~2t/ha/yr extrapolated) are possible using the daily manure output from two draught buffaloes alone. Such animals are fed only rice straw and allowed to graze poor quality forages. Fresh manure was added daily to 200 m^2 ponds and no other inputs were given. Ponds failed to develop blue-green algal blooms in all cases. Preliminary results suggest that supplementation with a cheap supplementary feed (e.g. rice bran) improves yields considerably (unpublished data, Asian Institute of Technology). Alternatively, supplementation of the manure with other nutrients (inorganic or organic) would improve natural food, and thus fish production.

Confinement and waste collection

Edwards (1983) has proposed integrated ruminant and fish culture for small-scale farmers in North East Thailand. The collection of adequate fresh buffalo wastes, which are bulky and heavy, from the field may deter their use in small-scale fish culture. If the buffaloes are confined near the pond this is less of a problem, although this would reduce the time available for grazing.

Sometimes buffaloes may be confined for longer periods such as just prior to and after the birth of young and at such time complete collection of wastes may be made possible (Cockrill, 1974). Where the animal is particularly valued for its production of milk, and yield improvements are sought by supplementary feeding, increased periods of confinement may also occur.

Buffalo platforms, where the animals are penned on rafts in marshy areas, have been used in South Kalimantan, Indonesia (Cockrill, 1974) and this system may be more widely applicable for confining buffalo directly over fish ponds.

4.3.5 Goats and sheep/fish culture

General considerations

The small size of these animals, their early maturity and low capital investment per head are particularly advantageous for small-scale farmers when compared with the large ruminants. Goats and sheep are found to play a significant role in many types of small farm system in Asia, Africa and Latin America. In Asia, over 60% of the sheep and goat population is raised on farms up to five hectares in size and these ruminants therefore could be used to improve the productivity of small-scale fish culture practical on this size of land holding. Details of goat and sheep production in the tropics are given in an excellent treatment by Devendra and McLeroy (1982).

The importance of sheep and goats in many developing countries is now being recognised but this is not yet reflected in any widespread integration with fish culture. Perhaps this is not surprising when it is considered that largely because of these animals' ability to use poor pasture and foliage, one of their major advantages is that they are readily stocked on very extensive pasture in dry, badly watered regions where fish culture is more problematic and has no tradition. More intensive husbandry systems do exist however, especially in Asia. Devendra and McLeroy (1982) cite the confinement of goats fed primarily with irrigated pangola grass (*Digitaria decumbens*) in Jamaica, guinea grass (*Panicum maximum*) in Malaysia, and leaves of jackfruit (*Artocarpus integrifolia*) and cassava (*Manihot esculenta*) in Indonesia. The small individual size of goats and sheep can permit direct and physical integration by convenient stalling over or above the fish pond just as takes place with pigs.

Box 4.6. Estimating the potential for sheep and goats in fish culture.

Using a 1000 m² fish pond

	Sheep	Goats
Stocking rate	at 450/ha ~45/pond	at 400/ha ~40/pond
Feeding rate	feed grass @ 10 kg/day total – 450 kg/day feed soybean cake @ 4 kg/day ~180 kg/day	feed grass @ 10 kg/day total – 400 kg rice bran @ 5 kg, total = 200 kg cotton seed meal @ 5 kg total 200 kg
Wastes	output ~1·07% TLW ave wt of sheep of 30 kg ~14·5 kg/day wet wastes	output ~1·1% TLW ave wt of goat of 40 kg ~17·6 kg/day wet wastes

To estimate potential fish production, assume a FCR similar to that of cattle wastes, on a similar dry matter basis (~35%)

Thus DM of wastes is 5-6 kg/day
Fish weight gain @ 3:1 is 1·6–2·0 kg/day
On a per hectare basis, ~16-20 kg/day, or 5·5–7·0 t/ha/yr

N.B. TLW = Total live weight

Sheep/fish culture

Sheep/fish systems in Java have been described, where the principal fish cultured is the herbivorous giant gourami (*Osphronemus goramy*); when stocked very densely at approximately 30 fish/m^2 high yields of small fish were obtained (Djajadiredja *et al.*, 1980). Sheep, stocked at a ratio of approximately 475/ha were housed over the pond in simple sheds, and fed grasses (10 kg/sheep) and soybean cake wastes (4 kg/sheep) on a daily basis.

Goat/fish culture

Goats are perhaps more suitable for intensive production than sheep. Goats are primarily browsing animals which make poor use of pastures but can efficiently utilise cut forage. The large variety of feeds and fodders that can be utilised by goats (Appendices 7 and 8) indicates their inherent flexibility within the agricultural system.

No information on prospective stocking ratios of goats per area of fish pond has been found, but judging from those of cattle and sheep integration, 400-500/ha would be feasible. The wastes from the few goats typically raised by farmers, if they were the only input, would only therefore be adequate for small pond units (e.g. 10 goats/200 m^2 pond).

4.4 Other potential livestock/fish culture

4.4.1 Rabbits

Integration of rabbit production with fish culture has not received much attention, but seems to have considerable potential. Large-scale, intensive produc-

Box 4.7. A check list for use of animal wastes in fish culture.

Type of animal, and fish?
Consider religious and social taboos; demand for animals and/or their products, marketing difficulties, investment cost and farmer's attitude to risk.

Production cycle of animal, and fish?
Consider waste availability and requirements for fish, changing labour requirements, changing climate and ability of pond to use wastes.

Feeding regime of animals?
Quantity and quality of feed to animals; amount of waste feed available to fish, frequency of feeding, animal feeds grown on the farm or purchased.

Are the wastes collected or can the animals be housed near the ponds?
Consider cost and labour requirements for transporting wastes.

Are the animal wastes fresh, diluted, well rotted?

Is some form of processing an advantage before use?

Will the wastes be sufficient to optimise fish production? If not are other fertiliser or feed supplements available?

Effects of changing production of animal or fish on the management and profitability of the other?

tion of rabbits is not common, although units of up to 10000 breeding females have been reported in Europe. Such units tend to rely heavily on concentrates as feed, which negates a crucial advantage of rabbits as a potential source of meat, as they can be fed much higher levels of roughages than pigs and poultry and are thus less competitive for high quality grain feeds (Swick *et al.*, 1978). In many ways rabbit, if locally marketable, is the ideal animal for integration with small-scale fish culture. Confinement can be in a similar manner to that used for chickens, over the pond in simple bamboo or wood cages. Pilot-scale rabbit production in North East Thailand suggests such husbandry practices may also reduce juvenile mortality and improve male fertility (often a problem in hot climates). Disease problems that are common in large-scale intensive production might also be reduced in small-scale ventures.

Analysis of rabbit manure (Elemele *et al.*, 1980) indicates that it may have greater value as a direct food for fish compared to other livestock wastes. Certainly, the pellet size and semi-floating nature of rabbit manure encourages enthusiastic and direct ingestion by fish. The proportion of true protein, compared to non-protein nitrogen is high in rabbit manure as is its energy value compared to dried poultry waste, broiler and cattle wastes.

CHAPTER 5
PLANTS IN INTEGRATED FARMING

5.1 General considerations

Plant materials of both wild and cultivated origin are commonly used as feeds for fish in ponds. Whilst there are limitations to the inclusion of plant materials in complete feeds for intensively raised fish, they can be very useful as supplementary feeds in more extensive systems.

In China, pond fish culture relies heavily on the direct feeding of green fodders to the macrophagous grass carp and the Wuchang bream (*Megalobrama amblycephala*). In some instances plant materials may be added to ponds without stocking herbivorous macrophyte feeding fish. In such cases they may act more as 'green manures', by way of the vegetation decomposing and acting as a fertiliser, than as a direct feed. The use of 'tatsao', a mixture of young, soft plants that decay easily, is widespread in China and Eastern Europe, especially in nursery ponds.

Plant materials are also used as organic fertilisers during the preparation of the pond prior to filling. Ark (1959) used aquatic weeds in such a way by piling them up and allowing them to rot before applying them to the pond. Martyshev (1983) described the growing of green manure crops in ponds before filling with water for fish culture. Combined crops of grains and legumes, where the 'tops' were fed to cattle and the remainder left for the fish were typical. Both coarse and soft materials were recommended for use at rates of 1-4 t/ha.

When herbivorous fish are present within a polyculture much of the benefit from green fodders may be obtained from the large amounts of faeces that they produce.

The feeding of plant materials is particularly relevant when animal manures are not available in sufficient quantities, either through a shortage of livestock or through competition with other users. In such cases heavy supplemental feeding of plant materials to herbivorous fish in the polyculture will produce manure within the pond, an ideal method of waste addition. The grass carp has become known within China, where in some areas animal manures are not heavily used, as a 'living manuring machine'. Macrophagous tilapiine fish, such as *Tilapia rendalli* and *Tilapia zillii* may also fulfill a similar function.

In order to feed green fodder effectively to plant-eating fish, it is necessary to understand both their feeding habits and feeding efficiency. Thus, Edwards (1982[b]) has classified potential candidate species of fish as:

— voraciously herbivorous
— omnivorous, consuming some plants directly but more especially after some decomposition and
— fish that have yet to have their potential established (Table 5.1).

Concerning green foliage in general, cutting when young and using fresh gives a feed of high quality, in terms of both crude protein content and palatability. Succulence appears to be a major factor determining preference but taste also appears to be involved (Alabaster and Scott, 1967). Grass carp, for instance, were found to prefer more fibrous plants over the succulent watercress (*Nasturtium officinale*) (Cross, 1969).

Environmental factors are also known to affect the feeding of herbivorous fish. Thus, feeding intensity in grass carp has been shown to increase with temperature, also feeding is found to become less selective (Alabaster and Scott, 1967). Feeding preference has also been found to alter with age (Mehta *et al.*, 1976).

The use of plant materials in fish culture has considerable potential where field cropping generates an excess of crop residues and by-products. By nature these tend to have a low economic value, and high bulk. Alternative uses may be hard to find, and local fish production may be ideal for using these materials locally.

Many plants may be made more suitable for use as pond inputs by some form of processing to improve palatability and food quality, and this is discussed in section 9. Finally, even where the local agricultural system is badly impoverished, cultivated or wild plant material may still be freely available to fuel an integrated fish culture system.

Plant materials used as fish pond inputs can be of terrestrial or aquatic origin. They may be grown specifically for use as fish feeds or used as or when they become available or gathered from the wild. Sometimes plant residues are used as pond inputs, the major crop being reserved from direct human or animal feed. Plants may also have a dual role, for instance, to stabilise pond embankments and also to provide feed for the fish. Integration of aquatic plants with fish has yet another dimension—the control of 'weed' plants, principally in irrigation systems serving agriculture, by the release of herbivorous fish into the system.

Observation of the fishes' preference for certain plants (see below) and their feeding rate is best made by using a feeding circle or enclosure. Generally made of rattan or bamboo, the enclosure will stop plants drifting across the surface of the pond and concentrate the feeding where it is easily seen. Also, residues under the enclosure can be examined for evidence of over or under-feeding, feed preference and decomposition rate.

5.2 Terrestrial plants

5.2.1 Purpose-grown

In China, crops have often been planted exclusively as fodder for herbivorous fish (FAO, 1977[b]). Typically grown on the wide dykes between ponds, both grasses and vegetables were cultivated.

Edwards (1982[a]) reported the use of elephant or Napier grass (*Pennisetum* spp.) in the Pearl River area, and English rye (*Lolium perenne*) and Sudan grasses (*Sorghum sudanense*) in the Yangtze River area. They would normally be cut repeatedly, fertilised and irrigated with pond sediments and water. Worthwhile yields of Napier grass (62·5 t/ha) have also been obtained for use as green fodder under much poorer conditions in Malaysia (Hickling, 1962), than those in China (225 t/ha, FAO, 1977[b]). Waste land was covered with excavated acidic soil (pH4), ridged up, and dosed with limestone and small amounts of fertiliser before being planted with stem cuttings. Other grasses purpose-grown for fodder include guinea grass (*Panicum maximum*) in Thailand, and lalang grass (*Imperata rundinacea*) in Malaysia. Grasses, in particular, can also perform the useful function of pond bank stabilisation, as well as being a source of fodder. In North East Thailand, pasture grass mixtures with legumes, such as centro (*Centrosema pubescens*) and stylo (*Stylosanthes humilis*) have both been used with success. Tropical kudzu or puero (*Pueraria phaseoloides*) is a widespread, creeping vine used in the same way. The vine has deep roots and is not affected by short dry spells. The leaves and stems will also spread out over the water and can be browsed directly by fish (Hickling, 1962).

Star grass (*Cynodon plectostachyus*) has also been used to good effect in both Africa and Asia for pond and dam stabilisation. It can also remain productive under dry conditions, an advantage for fish culture in regions of seasonal rainfall.

Box 5.1. Assessing potential of growing and feeding grass to herbivorous fish.

Perhaps the simplest approach is to use appropriate food conversion ratios, which for most plant materials are in the range 30-40:1.

Thus an annual production of 3000 kg/ha of fish will require up to 3000 × 40 – 120 t plant material (~ 330 kg/day). In China, this could theoretically be grown on 0·5 ha of land. With a typical pond of 2000 m^2 say 40 m × 50 m with 1·5 m borders and side slopes, perimeter area is 270 m^2, which could support say 1·6 t annual Napier grass production, hence 40 kg fish (200 kg/ha). Thus to support grass carp production of 600 kg/ha only from grass grown in the vicinity of the pond, a margin 30 m wide around the pond must be planted and cultivated intensively.

A more accurate approach is to consider digestible nitrogen content, and the actual acceptability to the fish; does it have to be chopped first? Is it of fertiliser value only?

Zweig (1984[b]) reported the culture of barnyard grass (*Echinochiao crusgalli*) in dry nursery ponds prior to their use for raising Chinese carp fry. The grass seed, available as a weed from rice paddies was sown at a rate of 75-100 kg/ha and after 45-60 days of growth, flooded. Decomposition then occurred for 7-10 days before fish fry release, by which time an abundance of zooplankton had developed and dissolved oxygen had stabilised.

The Chinese also grow a variety of vegetables for feeding fish, particularly the cabbage family (*Brassica* spp.) (Edwards, 1982[a]). Ruddle (1985[a]) also reported the growing of maize (*Zea mays*), sorghum (*Sorghum* spp.) and sweet potato (*Ipomea batatas*) for feeding fish.

5.2.2 Opportunist use

The fish pond provides a useful means of using spoilt and reduced value crops. If these products are only available irregularly, recycling them through other less flexible livestock may not be worthwhile.

Cash cropping is often a high risk business, and the small farmer is especially subject to market fluctuations. When prices are depressed the alternative feeding of some or all of the crop (e.g. maize, cassava) through the fish may reduce this risk. Also, where crops are grown on the farms specifically for livestock consumption, advantage may be taken of the economies of scale and extra production may also be used to feed some to fish. Maize silage is used as the basis for pig diets in Hungary, for instance, and is also fed, conveniently close to its site of production, to fish. Hepher and Pruginin (1981) have noted that heavily fertilised ponds are rich in protein, and that inputs of food energy limits fish production; hence the use in Israel of energy rich grains such as sorghum (*Sorghum durra*) as supplemental feeds. On the basis of yields from small integrated ponds in North East Thailand, Edwards (1983) also suggested that the use of energy rich rice bran and broken rice compounds does much to improve yields in manure-fed ponds. Crocker (1983) reported that a surprisingly high proportion (35-40% on a dry matter basis) of 'fine feed' i.e. grain was used in some ponds in China, where fish production is a specialisation. Where crop production is on a large scale, the regular and accepted losses of produce (harvesting and storage residues) may be used successfully for feeding fish. For instance, in Zambia, spilt soybean (*Glycine max*) was used as part of a complete diet for *Oreochromis andersoni* whilst De Bout (1955) noted the use of sweepings from corn mills were the food basis of high fish yields in the Katanga area of Zaire.

Roadside grasses and vegetation can also be an abundant additional source of fresh fodder for fish and other animals. Earthen storm drains, that also receive domestic runoff and wastes, can be rich sources of both duckweeds, and protein rich grasses such as para grass (*Brachiaria mutica*), even during a prolonged dry season.

5.2.3 Plant crop residues

Most crops grown for cash and subsistence purposes, have residues that can be recycled on the farm. In small quantities, the residues—whether vines, leaves or shoots—are often fed to livestock. When generated in large quantities these materials are frequently wasted, despite their usefulness as fish pond inputs.

In many cases the crop residue has a greater nutritional value than the harvested part. This is particularly the case with tubers and roots, which typically have very low protein and mineral contents compared with vines or aerial shoots.

The leaves of cassava or manioc (*Manihot esculenta*) are often used as a green fodder for fish. Analysis shows that even older leaves have a high nutrient value although, as with most green feeds, younger leaves contain more protein and less fibre, fat and ash. Mathieu (quoted by Hickling, 1962) also found them to contain valuable quantities of vitamins. The leaves of cassava have been fed to the fish without the stems and petioles, as these were not consumed by the fish (Hickling, 1962). Sweet potato (*Ipomea batatas*) leaves are also readily taken by fish, although livestock may compete for them as feeds. Young leaves from sugar cane (*Saccharum officinarum*) grown on the pond banks are frequently fed to fish in southern China (Ruddle, 1985[a]). McLarney (1978) has reported that the leaves of *Alocasia macrorhiza* have been found useful for feeding *Tilapia rendalli* in cages (Table 7.1). When stocked in ponds at 10000/ha (4-11 g), this fish was reported to reach over 300 g individual weight in 240 days when fed with *Alocasia macrorhiza* leaves *ad libitum;* unwanted recruitment was controlled with the predator *Cichla ocelearis* (Ramos-Henao and Corredor, 1980). *Cridosculus chayamansa* leaves (crude protein 24·2% dry matter) and *Xanthophyllum* spp. have also been recommended as herbivorous fish feeds.

The vines of the velvet bean (*Mucana* spp.) have been used as fish feeds in Malaysia and Zaire. These are mostly grown with maize, pearl millet or sorghum as a support (Gohl, 1980). Maize leaves themselves have also been used as feed for herbivorous tilapias (Hickling, 1962). Many leguminous plant products have been identified recently as potential feeds for fish by FAO (Appendix 9). Other residues are obtainable from perennial plants. The leaves of the papaya (*Carica papaya*) have been found palatable, as have those of the ramie (*Bochmeria nivia*), a crop grown mainly for the fibre contained in the stem. Canna (*Canna edulis*), is a perennial herb grown in the tropics, of which the leaves may be used. Banana leaves (*Musa* spp.) are generally considered a poor feed, but are nevertheless used when other forages are scarce. Leucaena, or ipil-ipil (*Leucaena leucocephala*) is a leguminous tree increasingly grown for timber and fodder purposes throughout the tropics. Both seeds and leaves contain appreciable quantities of high quality protein. Unfortunately, it also contains a toxic amino acid, mimosine, which inhibits growth even when incorporated as a dried meal into pelleted complete feeds. Both Jackson *et al.* (1982) and Cruz and Laudencia (1978) found a poor growth response from incorporation of this legume in diets; but in contrast some authors claim improved growth performance with its inclusion (Pantastico and Baldia, 1979, 1980). Wang (1985) reduced the toxic mimosine content by soaking the fresh leaves in water at 30°C for 48 hours. When incorporated in a complete feed this soaked leaf meal gave a significantly better growth response than either sun-dried (for two days) or commercial leaf meal. Thus, presoaking probably improves the quality of the fresh leaf as a feed. Other common leguminous tree products may be more useful, however, e.g. the leaves of sesbania (*Sesbania grandiflora*) and gliricidia (*Gliricidia maculata*).

The ability of perennial bushes and trees to provide fresh fodder throughout a dry season can be important for maintaining fish production year round.

Potential

Many of the residues traditionally used as feed for animals, particularly ruminants, may have application as pond inputs. Cereal straws, maize stover, sugar cane leaves, etc. have been used. Normally such materials would be piled up in the pond to attract fish as decomposition occurred. The highly fibrous nature of these products make them poor and unpalatable direct feeds. Indeed, it may be more beneficial to feed coarser plant residues through ruminants, and use the animals' wastes for pond inputs (see section 3). Other alternatives include some form of processing (see section 9). An economic assessment of low value/high bulk plant and plant residues as pond inputs is given in section 11.2.2.

5.3 Aquatic macrophytes

5.3.1 General considerations

Aquatic macrophytes, as distinct from phytoplankton, are considered as those higher plants which grow in water continuously. They may include those present in soils covered with water during a major part of the growing season. The abundance of aquatic macrophytes in many parts of the world has created a dilemma. Often considered as a weed problem to be controlled, possibly using fish, aquatic macrophytes can in suitable circumstances be viewed as a highly productive crop requiring no tillage, seed or fertilisation (Ruskin and Shipley, 1976). Edwards (1980[c]) has reviewed the pathways in which aquatic macrophytes may be used in food production, and has detailed potential integration with fish culture. Fish show preference for different types of aquatic macrophytes based on succulence and taste, as with plant materials in general, and this will affect their usefulness in weed control. Least favoured by grass carp, for instance, are the rushes, sedges, watercress, water lettuce and water hyacinth. Most favoured include the filamentous algae, soft submerged macrophytes and duckweeds.

The use of aquatic macrophytes as pond inputs need not be limited to feeding the more succulent varieties to strictly herbivorous fish. The use of water hyacinth for the Nile tilapia has been reported by Edwards (1985[c]). In this case, little direct ingestion by the fish of the fresh chopped weed seems to have taken place; rather the considerable yields (5-6 t/ha/yr extrapolated) are obtained via an enriched food chain in the pond.

Other potential uses for macrophytes broadly include their use as fodder for livestock, recycling wastes, as fertilisers and as direct human food.

5.3.2 Use of aquatic macrophytes in pond systems

(a) Culture of food plants and fish in the same pond

The culture of either submersed or floating aquatic macrophytes for feed with herbivorous fish together in the same pond is not normally an effective approach. Fertilisation of the pond to encourage the necessary growth of submerged macrophytes, would rather tend to favour phytoplankton growth. In turn, the submerged macrophytes would be eliminated by their shading effect (Edwards, 1980[b]).

Usually, floating macrophytes cannot be maintained in balance with a population of growing herbivorous fish. Conversely, if a lower stocking density of fish is used, problems may arise with the macrophytes covering the pond surface excessively.

(b) Cultivated aquatic macrophytes

The Chinese cultivate useful water plants in rivers and canals adjacent to fish farms, and they may form a major source of feed for both livestock and fish. Three aquatic plants in particular are grown, *Eichhornia crassipes* (water hyacinth), *Pistia stratiotes* (water lettuce) and *Alternanthera phylloxeroides* (alligator weed) (Edwards 1982[b]).

Duckweeds may also be conveniently grown in small ponds, harvested by skimming or seining with a net, and fed to herbivorous fish. Duckweeds are grown in shallow ponds converted from rice paddies on the outskirts of some urban communities in Taiwan. 'Blackwater'—basically domestic and human

Table 5.2. Productivities of selected aquatic macrophytes.

Species	Conditions	Yield (extrapolated) t/ha/yr	Reference
Water hyacinth	Artificial fertilised ponds	75·6–191·1	Yount and Crossman, 1970
	Fertilised ponds	70·8	Boyd, 1976
	Fertilised with sewage effluent	219–657	Wolverton and McDonald, 1979
	Grown in irrigation canals in China	400–750	Coche, 1980
Duckweeds	Average figure under tropical conditions	22·1	Edwards, 1980[a] (from other authors)
Lemna, Wolffia, Azolla	Grown in irrigation canals in China	150–187	Coche, 1980

wastes—is diverted from open ditches into the ponds as fertiliser. The duckweed is harvested weekly; ponds are skimmed with bamboo poles and the concentrated duckweed sieved into sacks and sold fresh, for carp fry feed. Importantly, only a proportion of the duckweed is removed at one time. A more or less complete cover of the duckweed must be maintained, otherwise the pond becomes dominated by phytoplankton and the duckweed diminish (Chao, 1983). Van Dyke and Sutton (1977) report efficient digestion of *Lemna* spp. fed to grass carp, and it appears to be a favoured food. Apart from high productivity (Table 5.2), duckweeds have favourable protein, fat and fibre contents, particularly when cultured in nutrient rich waters. In Indonesia, *Lemna minor* was found more suitable than the coarser *Hydrilla verticillata* and *Chara* spp. for feeding to the Nile tilapia in cages (Rifai, 1979, 1980).

The residues of aquatic macrophytes grown principally for human consumption can also be useful pond inputs. Taro (*Colocasia esculenta*), an emergent aquatic plant grown primarily for its starch filled corm, produces leaves valued as fish feeds. It is claimed that both the giant gourami (*Osphronemus goramy*), a vegetarian fish of South East Asia, and the plant-eating tilapias prefer it to the more commonly used cassava leaf. However, its nutritional content appears inferior, with more water and protein but less carbohydrate and fat. Water spinach (*Ipomoea aquatica*) is produced in large quantities throughout Asia. Around Bangkok, in Thailand, it is often cultivated as a cover crop around fish ponds. The growing tip, with fresh young leaves and stem, is cut for human consumption and the remainder is thrown into the nearby fish pond as a fish feed.

Box 5.2. How to grow tilapia on water hyacinth.

To produce a yield of 110 kg/fish/yr from a 200 m^2 pond containing Nile tilapia (5·5 t/ha/yr) use a loading rate of 4 kg DM a day of either

fresh, chopped (5 cm pieces) water hyacinth, or aerobically composted water hyacinth

that is

40 kg wet weight of chopped fresh water hyacinth
or
~8 kg weight of composted water hyacinth (assuming compost ~90% DM)

This water hyacinth can be grown in reasonably fertile water in the tropics to yield 100 t DM/ha/yr. To provide enough water hyacinth to grow 110 kg tilapia/yr from a 200 m^2 pond, a water area of 150 m^2 is required, possibly fertilised with sewage or other wastes unacceptable for direct use in fish culture (after Edwards, 1985[c]).

Rice (*Oryza sativa*), the most important crop plant in the world, is usually grown as an aquatic annual (Cook *et al.*, 1974). It also generates large quantities of straw as a useful residue—approximately 1 kg of straw for every kg of grain (Morgan Rees *et al.*, 1977). Used fresh it is not a feed readily taken, but after some form of processing its value to aquaculture may be of considerable importance, especially where other resources are lacking. In China, when other plant wastes are limited, rice straw is thrown into the fish pond as a feed/fertiliser.

The use of *Azolla pinnata,* the aquatic fern that has a symbiotic relationship with a blue-green alga (*Anabaena azolla*), as a fish feed has been reported. The fern has been cultivated in rice fields as a green manure, and also used to supplement the diets of pigs and poultry. Its rapid growth (the fern doubles its biomass in 3-10 days) and high protein content suggest a use for it in integrated fish culture. The growth response of the Nile tilapia fed directly on fresh azolla in plastic tanks was, however, poor (Pullin and Almazan, 1983). Proximate composition of the material used indicates the probable reason, although protein content is adequate, fibre is high and available carbohydrate very low. The latter factors will in general decrease the value of the plant as a direct feed, especially for fish which are not direct macrophyte feeders.

5.4 General comparison of terrestrial and aquatic plants as pond inputs

5.4.1 Quality

Feeding trials with livestock have shown that many aquatic macrophytes are inferior as feeds compared with more conventional land grown crops. This is not necessarily because of chemical composition; some aquatic plants have a crude protein (CP) content in excess of that for typical terrestrial fodder crops. Plants such as duckweed have crude protein values as high as 42·6% (Myers, 1977). Duckweeds also have a low fibre content compared to most other plant fodders, which is an important factor in its use as a productive feed. Also, levels of the amino acids methionine and lysine, usually lacking in plants, are comparable to those in animal products (Porath *et al.*, 1979).

Variations in both terrestrial and water plant quality can also be related to seasonality and environment (see Table 5.3). Seasonal variability is often more problematic in terrestrial crops. Whilst cassava leaf shows negligible variation in quality during wet and dry season, or through a period of two months growth, guinea grass (*Panicum maximum*) can vary in crude protein (CP) by a factor of four (Gohl, 1980). In contrast, the nutrient content of the water can dramatically affect aquatic plant composition; a three fold increase in protein (up to 32% CP/DM) has been observed for water hyacinth in waste water compared to nutrient poor waters. However, the most critical difference between terrestrial and aquatic fodder crops is water content. The high water content of aquatic macrophytes, often exceeding 90% wet weight, poses a major constraint to their harvest and use in aquaculture. Terrestrial pasture grasses contain much less water (~80%) and therefore approximately only half as much is required to obtain the same dry matter input (Little and Henson, 1967).

Table 5.3. Range of protein content for selected green fodder crops.

Common Name	Species	Range of crude protein % (CP/DM)	Notes
	Typha latifolia	10·5–3·2	April–July
	Typha latifolia	4–11·9	Different sites
	Justicia americana	22·8–12·5	May–September
Water hyacinth	*Eichhornia crassipes*	9·2–4·7	Summer–spring
	Eichhornia crassipes	11·3–32·9	Nutrient content of water
Water spinach	*Ipomoea aquatica*	18·8–34·3	Whole plant, mid bloom, in Nigeria–fresh leaves and shoots, Fiji
Guinea grass	*Panicum maximum*	5·3–20·5	Fresh, early bloom, Tanzania–fresh, cut weekly, Malaysia
Azolla	*Azolla pinnata*	18–27	Bangkok strain in the Philippines
Napier, elephant grass	*Pennisetum purpureum*	10·8–5·9	Fresh, 4 weeks old, fresh, 12 weeks old –Thailand
Cassava, tapioca	*Manihot esculenta*	24·8–25·4	Fresh leaf, wet season, 4 weeks–fresh leaf, dry season, 8 weeks
Sweet potato	*Ipomea batatas*	19·4–21·9	Fresh leaves–Israel, fresh vines–Trinidad
Papaya	*Carica papaya*	20·9–32·6	Fresh leaves–Pakistan, fresh leaves–Nigeria

5.4.2 Food conversion and feeding rate

Fish production using plant materials is typified by high food conversion ratios (FCR). This is a direct result of their high water content, but also reflects the poor digestibility of many green fodders. The FCR of terrestrial Napier grass has been reported as 27:1 (Hickling, 1962) but because of their higher water content those of aquatic plants may be much higher. Edwards (1980[c]) has observed that FCR values quoted are often difficult to evaluate. Over or under feeding is a common problem that leads to error. Also, feeding trials are conducted under varying experimental and environmental conditions. A consequence of poor food conversion is the need for supplying large amounts of feed, and this may be the most serious constraint to using aquatic plants as feeds. Harvest and feeding in large ponds is very labour intensive. Thus, especially for large-scale pond culture of fish, the use of secondary processing to increase the food value of plant materials may be necessary.

5.4.3 Management of plants as pond inputs

Overfeeding, and accumulation of green fodders in the pond, can lead to oxygen stress. Plants, especially those that are fibrous or coarse, as shown by the COD/BOD ratio (see Table 5.4), decompose relatively slowly compared with unadulterated animal manures.

On this basis, Edwards (1982[b]) estimated that five days after addition to ponds only 14% of *Typha,* 18% of *Eichhornia* and 35% of *Najas* would have been oxidised. This infers that compared to animal waste-fed ponds, oxygen problems may occur more gradually (over a number of days) but be more sustained unless the accumulated decaying plants are removed. More succulent, less fibrous plants, rot faster (lower COD/BOD ratio) than older and more fibrous types and cause less problems in this respect. The data listed refers to aquatic macrophytes, although terrestrial fodders would have oxygen demands in the same range (Edwards, 1982[b]). The lengthy period for complete decomposition of plant materials results in a build-up of organic sediments on the pond bottom. In China, these sediments are laboriously removed for fertilising the dykes nearby, and are the basis for the high yields of mulberry and vegetables achieved.

Box 5.3. A checklist for use of plant products in fish culture.

Type of plant product: high value (grain/by-product)
low value (coarse vegetation)

Use primarily as a feed or fertiliser?
If used as a feed (1) are herbivorous fish available, marketable?
(2) approximately what FCR can be expected?

Quality of feed? Palatability, dry matter, digestable protein, fibre and lignin content.

How much material is required for a given fish yield? Are such quantities conveniently available (consider labour requirements)?

Should the material be processed before use? Chopped, fermented, composted?

Is it worth feeding the material through another animal first (before using the animals manure for fish)?

If using a polyculture based on feeding plant roughage to herbivorous fish, are other suitable fish available to complement them?

Table 5.4. Chemical oxygen demand (COD) and biochemical oxygen demand (BOD) of aquatic macrophytes compared with animal wastes. (After Edwards, 1982[a], modified from Almazan and Boyd, 1978 and Taiganides, 1977.)

Aquatic macrophyte	Life form	COD (mg O_2/mg dry weight)	BOD (mg O_2/mg dry weight/day)						Ratio of COD / BOD_5
			$BOD_{0\cdot5}$	BOD_1	BOD_2	BOD_3	BOD_4	BOD_5	
Typha latifolia	Emergent	1·08	0·045	0·067	0·105	0·121	0·145	0·148	7·30
Eichhornia crassipes	Floating	0·88	0·044	0·079	0·109	0·128	0·145	0·157	5·61
Najas guadalupensis	Submersed	1·09	0·077	0·107	0·187	0·283	0·313	0·383	2·85
Mean		1·02	0·055	0·084	0·134	0·177	0·201	0·229	5·25
Pork pig waste									3·3
Laying hens waste									4·3

N.B. All figures given for 30°C

CHAPTER 6
FISH/RICE SYSTEMS

6.1 General considerations

Fish culture can be integrated with agriculture in ways other than utilising waste or by-products. Growing fish on the same piece of land as other aquatic crops, especially rice, can bring advantages to both crops whilst intensifying land and water use.

Rice and fish are considered complementary and basic foods in South East Asia, where the practice of fish-in-rice culture is most developed. The potential for its development elsewhere, however, is clear in that rice constitutes the world's major food source, using 50% of world's arable land (Swaminathan, 1984). Other aquatic macrophytes, however, are far less widely grown but may be of considerable importance locally. They will often have potential for contributing to increased and diversified food supply, especially if their production is integrated with fish culture.

In many areas fish culture in ponds may not be feasible for reasons such as lack of land tenure or high capital cost and risk (see Section 11.5.1). In such cases stocking of fish with aquatic crops may be the only practical alternative permitting fish production and increasing returns on the use of land.

Intensification of rice growing practices, particularly the introduction of high yielding varieties (HYV) of rice, has caused a decline in fish/rice culture mostly through effect of pesticides on the growth and survival of fish. In the longer term, the development and use of pesticides less toxic to fish, together with a greater understanding of biological pest control, should further stimulate fish-in-rice culture. Also, there is great potential in growing fish with other aquatic crops such as lotus (*Nelumbo nucifera*) and water chestnut (*Trapa* sp.).

It is accepted that the stocking of fish in paddy fields more effectively uses the available food and space. This is particularly true if a polyculture of fish is used. The stocking of fish to feed on competing phytoplankton and submerged weeds benefits the rice, particularly if done at the start of rice production. The stocking of macrophagous tilapias (either *T. zillii* or *T. rendalli*) within one week of transplanting rice, was claimed by Mortimer (quoted by Pullin, 1983) to control weeds, while stocking 5-9 weeks after transplanting was not effective. The activity of feeding fish and their excretion can also improve paddy fertility to the benefit of rice production. Recent research in China suggests that this fertilisation effect is more marked in less fertile fields, and can be cumulative over a few years (Spiller, 1985).

Hora and Pillay (1962) reported that the yield of rice increased approximately 15%, on average, in the Indo-Pacific countries when fish are introduced. Khoo and Tan (1980) reported that in China and Russia, increases in rice yields of up to 14% and 7% respectively have been recorded.

In some cases, though, a decrease in rice yield has been caused, mainly because of the unplanted areas given over to fish refuges within the paddy field, e.g. 15% in some areas of Indonesia (Ardiwinata, 1957). In the rice growing area of the Central Niger Delta however, Mathes (1978) reports quite severe depredations on rice by herbivorous fish. The fish, wild species entering the rice growing area freely, damage the crop by either eating the grain (*Alestes* spp.) or by consuming the stalks and leaves (e.g. *Distichodus* and *Tilapia zillii*).

ADVANTAGES

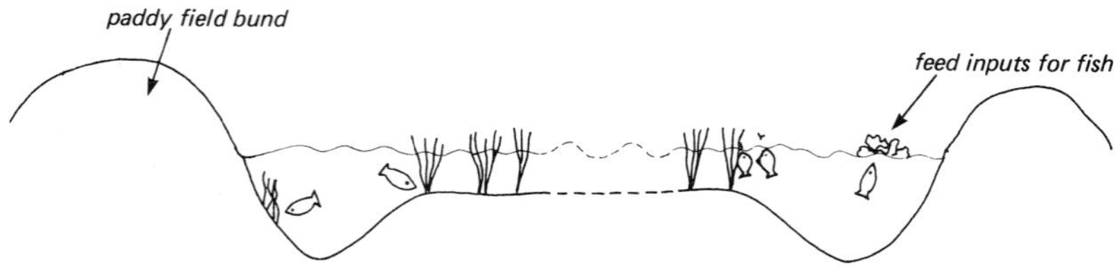

Weed Control by herbivorous fish

Feed Inputs for fish also act to improve fertility for rice production via feed residues and wastes

Labour- minimum extra labour required

Income- if plant is long maturing,simultaneous fish culture may provide an earlier financial return

Land- efficient production of high carbohydrate(rice)and high protein (fish) food from same area of land

Water-higher level of water management for fish may improve conditions for plant crop

DISADVANTAGES

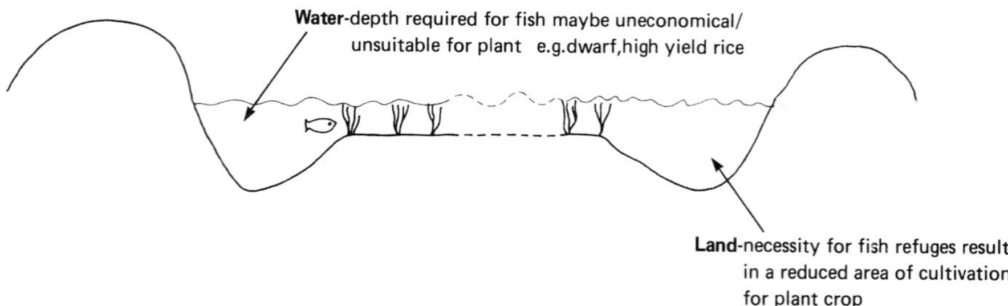

Damage to plant crop by herbivorous fish

Fertiliser demand may be increased in order to stimulate natural food production for fish

Labour- demands may be too high,especially where plant crop production is very mechanised

Pest Control- timely and efficient use of pesticides may be difficult without risk to fish

Fig. 6.1. Summary of the main advantages and disadvantages of raising fish with other aquatic crops.

The vulnerability of HYVs to attack by diseases and pests is well known by farmer and researcher alike. The stocking of fish in paddy fields may affect conventional pest control in rice in two basic ways. The farmer may simply reduce or stop using pesticides, with consequent increased risk to rice production or revert to local, more pest resistant varieties of rice. The use of HYVs by farmers is subject to the seeds, fertilisers and pesticides being available at the right time and price, but may also be affected by other constraints (Grandstaff and Grandstaff, 1986). If pesticide use is continued, the risk to fish culture may be reduced however by following certain guidelines (see section 7.5). The development of rice varieties that are more disease resistant and a trend to pesticides less harmful to fish are optimistic signs for more widespread fish-in-rice culture in the future (Spiller, 1985). Pullin (1985) reported the research effort into organic fertilisers, such as azolla, for rice fields and natural pesticides less harmful to fish. Neem oil and cake, derived from seeds of the neem tree *Azadirachta indica* have shown potential and are widely grown in India and Africa. In addition the benefits of fish as a biological control of rice pests are being acknowledged. Fish are known to control many of the common insect enemies of rice. Stem borers and plant and leaf hoppers, for instance, were found to decline by between 12-75% when fish were released in rice fields in China (quoted by Spiller, 1985). Basic advantages and disadvantages to fish/plant culture are shown in Fig. 6.1.

6.2 Methods of fish culture in rice paddies and ponds

The basic methods of rice/fish culture, which may also apply to the integration of fish with water spinach and other aquatic macrophytes, are described and compared in Table 6.1. The rearing of fish with rice is commonly grouped into:

— captural methods—in which indiginous stocks are trapped in the newly flooded fields, and
— cultural methods—in which stocks of fish are deliberately introduced.

Naturally occurring fish of economic value found in paddy fields tend to be slower growing and/or predatory, such as the snakehead (*Channa striata*) or climbing perch (*Anabas testudineus*). So-called cultural systems for raising fish-in-rice are therefore more likely to benefit the rice crop and provide a better production of fish than captural methods. Indeed, in the management of fish-in-paddy culture systems, every effort is usually made to remove the wild fish and prevent them from entering the culture area, to avoid predation on the more valued stocked fish. However, these fishes may be preferred for reasons of taste and value; in parts of North East Thailand where the introduction of cultured fish species has recently been attempted, the technique was only acceptable to farmers when one-way screens allowing the entry of wild fish were placed in rice paddies.

Cultural methods vary according to the climate, water availability, fish species, plant variety and traditional custom.

Table 6.1. Comparison of captural and cultural fish/rice culture.

	Captural	**Cultural**		
		Concurrent (Bersama)	Rotational *Palawidja*	Rotational *Panjelang*
Description	No stocking of fish, entirely of fish populations entering flooded ponds and paddies, where reproduction takes place. Harvest on fish entering, during and at end of crop growing season	Fish reared simultaneously with the growing of the plant crop	Fish cropped after a single annual rice crop	Fish cropped between harvest and next replanting
Occurrence	Very widespread compared to cultural, e.g. 4 million ha Indonesia	Hong Kong, India, Indonesia, Japan, Malaysia, Thailand, Vietnam	Indonesia, Thailand	Indonesia
Advantages	Cost: minimum inputs required some adaption of paddyfield	Fish increases plant production by fish culture residues, also help to control pests, weeds, insects, molluscs		Particularly useful for short term fry production May reduce water use
			Pesticides, fertilisers, modern intensified cropping method may be used. Reduced water needs, reduction of pesticide accumulation in fish Pest control by interrruption of life cycles Plant residues benefit fish (e.g. rice stubble, split grain) Fish residues benefit plant production (not as much as concurrent) Reduced production costs–since paddy bottom soft and clean after fish culture	
Disadvantages	Natural/wild fish are typically carnivorous, not most productive. Few benefits of fish culture to plant, fish not normally fed	Plants/fish may have different water requirements. Limits use of fertilisers and pesticides	Only possible where water available year round i.e. irrigation systems. Highly intensive cropping methods and profitability threaten viability.	

Spiller (1985) has classified fish/rice culture systems into:

— traditional low input/low output systems adapted to traditional rice varieties, and
— modern systems of fry and food fish production with higher management inputs and higher yields, which are adapted to the management requirements of modern varieties of rice.

In addition, fish culture in paddy fields may be concurrent with rice culture or in rotation (see Fig. 6.2).

Captural systems are often very important sources of fish where traditional rice growing methods are used. Importantly, they may, with adaptation, have potential for some intensification and subsequently higher yields. Attempts to upgrade these systems may be incremental however. The rice farmers attitudes often need to change from that of the hunter to fish farmer.

Production of fry, fingerlings and market sized fish are all practised in rice fields. Fry and fingerlings require a much shorter culture time and may be more suitable when rice production methods require regular draining of the paddy. Also, where rice production is almost year round, a crop of fry or fingerlings may be quickly taken in between, such as is done with the Panjelang system in Indonesia (Table 6.1). Climate and water availability may make the production of food fish difficult, depending on the accepted marketable size. Supplementary foods are commonly given in more intensive systems, especially when the culture period is limiting (see Table 6.2). Such feeding can allow fast enough individual growth to ensure that the required sized fish are produced. High levels of feeding can be counterproductive however. Supplementary feeding of rice bran and mustard oil cake (1:1) to common carp above the 4% level (body wt/day) was found to be wasteful and to cause a deterioration in water quality, in India (Ghosh *et al.*, 1984). In Japan, silkworm pupae, dried mysids and other small shrimps, and cereal bran were provided (Kuronuma, 1980), although the practice has now all but disappeared. In Thailand, organic manures and rice straw compost have been used in addition to rice bran. Chen (1954) suggests that 50-100% more fertiliser is required than in normal rice field practice; this additional input being required for primary production and the associated fish food chains.

6.2.1 Preparation of paddy fields

(a) *For rice*

According to Singh *et al.* (1980), for rice cultivation, either dry land or wet land preparation can be undertaken. In dry land preparation, the fields are thoroughly drained and the land is prepared with the use of heavy harrows, chisels or stubble mulch equipment to stir the soil without inverting or covering the stubble, so that weeds are allowed to dry out. In wet land preparation, the field is ploughed and harrowed crosswise and lengthwise until the soil is friable. Immediately after ploughing, more water is added to speed up the decomposition of weed and crop residues that are ploughed under. The land is tilled at

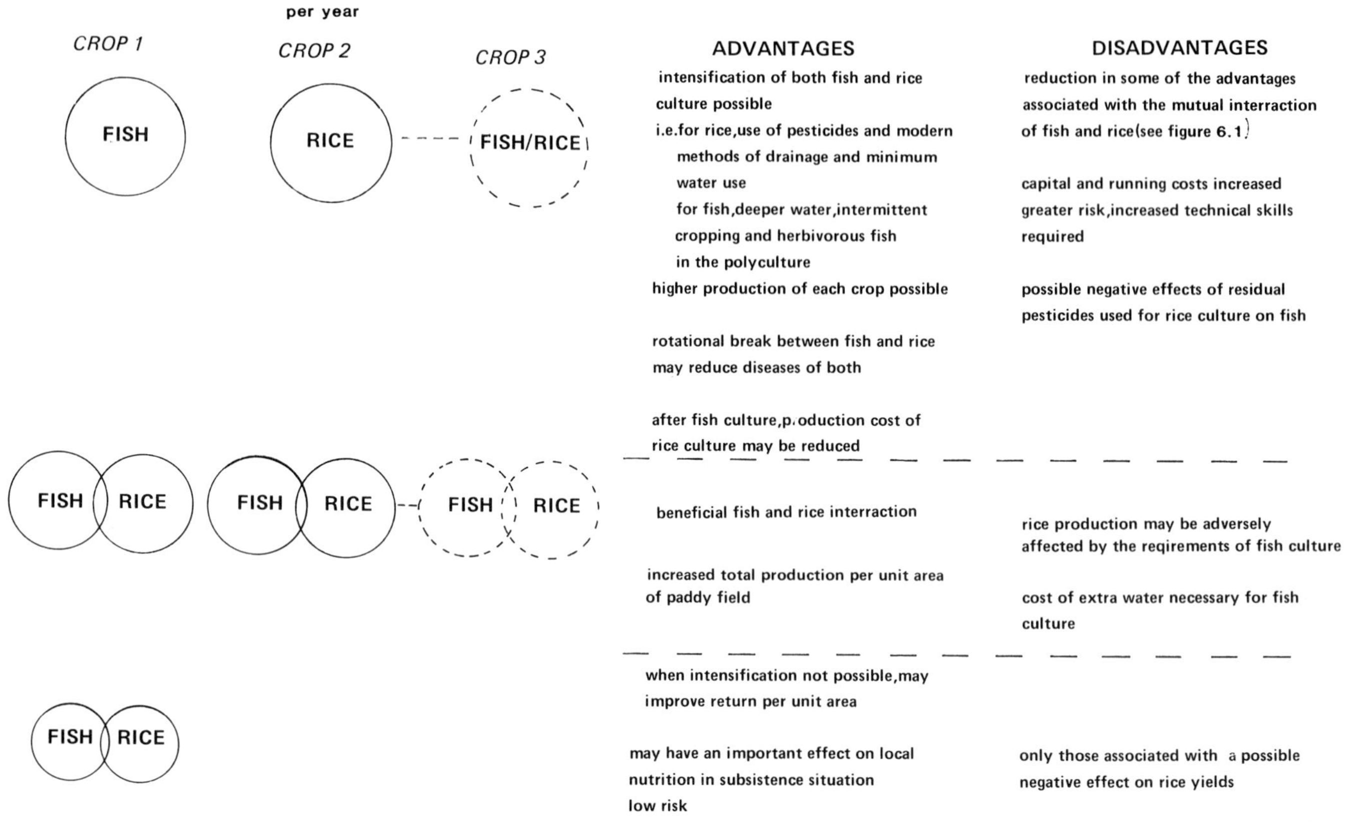

Fig. 6.2. Concurrent and rotational fish-in-rice culture.

Table 6.2. Comparison of yields for fish grown to table size in rice fields under different conditions.

Yields kg/ha		Notes	Reference
Monoculture	Polyculture		
78–220	130–293	No feeding, 69–109 day culture period. Concurrent culture, Philippines. *O. niloticus, O. mossambicus, C. carpio*	De la Cruz, 1980
50–70		*C. carpio* Culture period, 75 days, concurrent. Indonesia	
40–60		*Puntius goniotus*	
	60–100	*C. carpio* and *P. goniotus*	Khoo and Tan, 1980
26–126		*C. carpio* 60–75 days culture period.	
	117–199	*C. carpio* 60–75 days culture period. *Helostoma temmincki Osteochilus hasseltii*	Djajadiredja *et al*, 1980
171–589		*O. niloticus,* culture period 72 days. Supplementary feeding and fertilisation, rotational culture. High stocking density; 5–10 000 fish/ha, water depth 30–40 cm	
	423–694	*O. niloticus* – Philippines and *C. carpio*	De la Cruz, 1980

least 15 days before transplanting, so that loss of nitrogen released by the decomposing organic matter is minimised. The last harrowing is carried out a week after the second harrowing to allow germination of the weed seeds and further break down of soil clods.

(b) *For other aquatic plant crops* (see Table 6.3)
When these crops are grown in the more marginal, less easily managed situations, for example in undrainable borrow pits or swamps, integration with fish culture is likely to prove more difficult. Since many of these aquatic crops are not grown intensively, little is known of their optimum requirements for water, nutrients, etc. Pressure for the use of water resources will undoubtedly encourage the intensification of production, perhaps along the lines already used for water spinach, and Chinese water chestnuts. Integration with fish should further improve the productivity of such systems.

The adaptation of shallow rice paddy fields for fish culture can take many forms, but basically involves the creation of some deeper areas as refuges for fish to obtain shelter during periods of high temperature or temporary draining of the paddy field. The area of trench or refuge will vary with the situation but in general would not exceed 10-15% of the total area. Some farmers gradually improve and extend such trenches into small ponds, recognising that such changes allow greater stocking densities of fish and subsequent yields. Spiller (1985) has identified eight different systems for rice/fish culture in Thailand, and the structural adaptations necessary are also given (Table 6.4).

Table 6.4. Fish production by rice/fish culture system in Thailand. (Source: Spiller, 1985.)

	Period	Fish productivity (kg/ha/day)	Fish production per crop (kg/ha)
Rain-fed systems			
Trap pond	120–150 days	0·25–0·4	30–60
Open dike	120–150 days	1·0–1·2	120–180
Closed dike	150–210 days	1·2–1·4	180–300
Terraced	150–250 days	2·4–2·6	400–600
Irrigated systems			
Rotational	120–140 days	6·3–6·4	750–900
Concurrent synchronous	120–150 days	2·5–6·0	300–900
Integrated farm	120–150 days	1·5–2·4	180–360
Concurrent asynchronous	120–140 days	2·9–3·3	400
Deep water culture	180–240 days	5·0–7·5	900–1800

Trench construction and other drastic changes to rice paddies for fish culture may not however be possible or necessary. Ramsey (1983) has described the reluctance of farmers growing rice in Ifugao, in the Philippines, to construct trenches because of traditional beliefs. Adaptations to traditional harvest pits instead were suggested to allow fish-in-rice culture (Fig 6.3). Taylor *et al.* (1986) have studied the use of small refuge ponds (3 m diameter) shaded with bamboo thatch instead of trenches. The rice was cultivated leaving one in four rows empty, the so-called 'border' method; this allowed the fish to obtain access to all parts of the field through the unplanted rows.

Adaptation of rice paddies may actually be easier in poorly levelled or terraced paddies (Fig. 6.4). Also there may be benefits to rice production by way of improved water management and increased plantable area resulting from such modifications. This contrasts with the well-levelled site, where the area planted for rice inevitably falls as the paddy field is modified for fish culture.

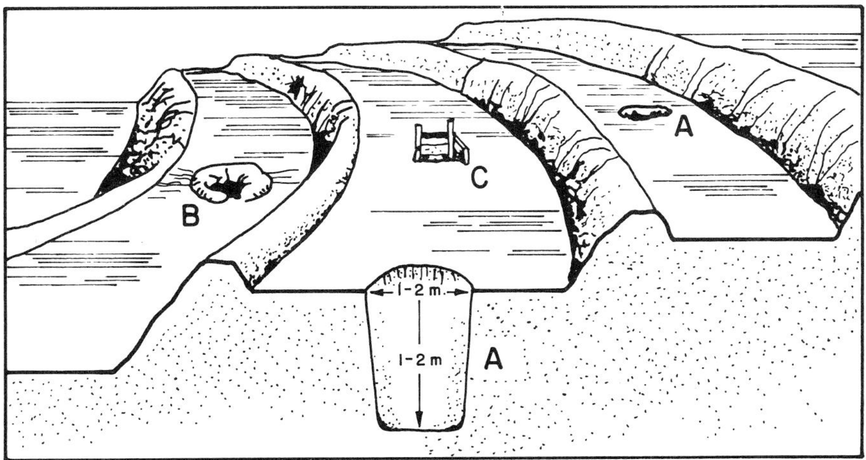

Cross-section of Ifugao rice terraces, showing traditional harvest pits, (A) standard pit, excavated from soil, (B) pit with secondary dike, (C) pit lined with wood to prevent erosion.

Fig. 6.3. Adaptation of rice fields for fish production using harvest pits or refuges. (After Ramsey, 1983.)

6.3 Species of fish utilised in plant/fish culture systems

Generally, fish grown in paddy fields are tolerant of shallow water (average depth approximately 15 cm), high and variable temperatures (up to 34°C and with 10°C range), and low oxygen concentration (Polich, 1979; Khoo and Tan, 1980). Table 6.5 shows the different species that are presently used in rice/fish culture operations.

The size of fry and/or fingerlings introduced into the paddy fields varies according to the fish species used, traditional practices and the location (Khoo and Tan, 1980). In Indonesia, the eggs of the giant gourami (*O. goramy*) are introduced into rice fields and fingerlings (~8-10 cm) are harvested and sold for stocking in other farmyard ponds (Khoo and Tan, 1980). Rice fields are now a major source of fry for stocking the on-growing systems using running water in Indonesia. Tilapia fingerlings of approximately 9-12 cm are stocked into rice fields in China (Tapiador *et al.*, 1977) and in Japan common carp fry and two year old fish have been stocked into rice paddies (Kuronuma, 1980). Crustaceans are also grown in rice paddy fields. The red swamp crayfish (*Procambarus clarkii*) has been unintentionally introduced into rice fields in the USA, and

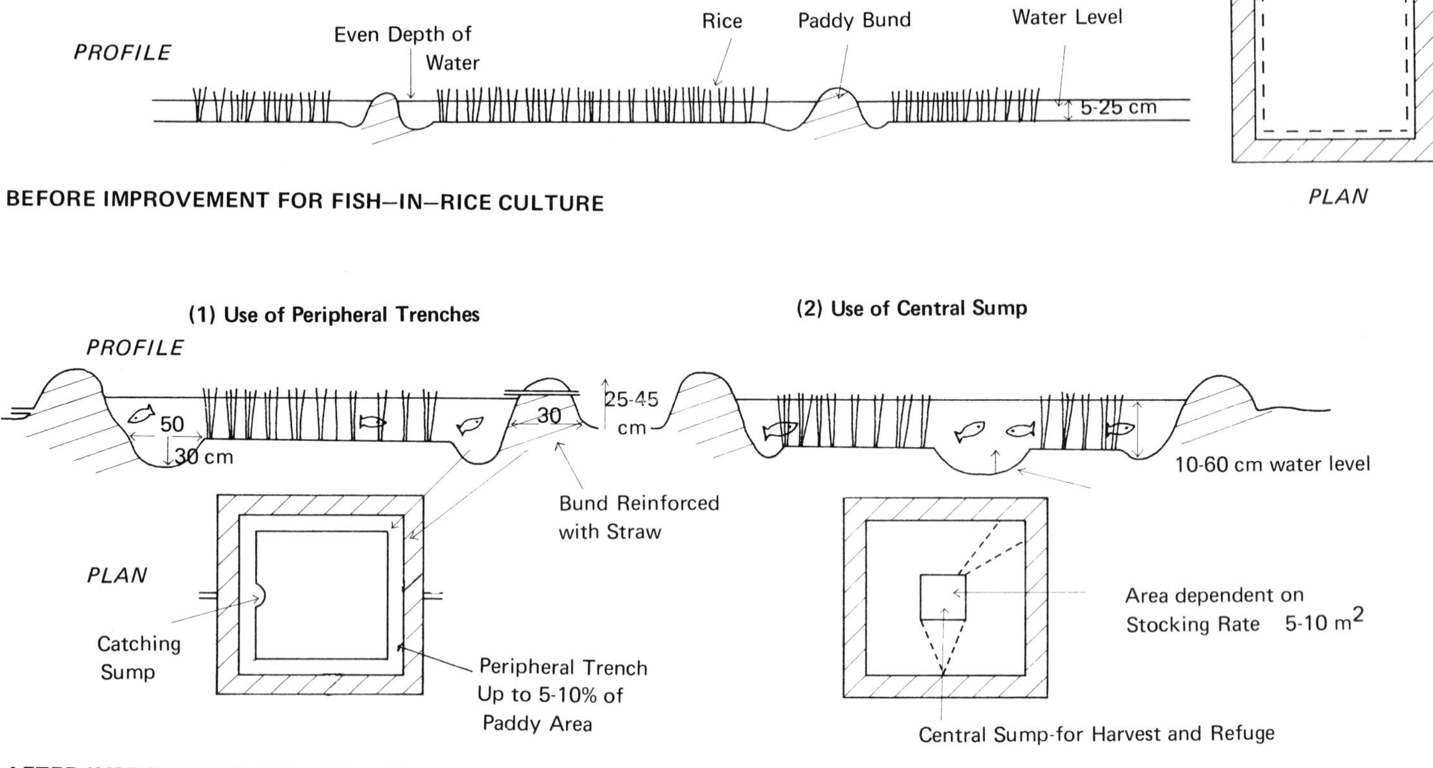

Fig. 6.4. Adaptation of rice paddies for fish production.

Table 6.5. Fish species used in rice/fish culture systems.

Scientific name	Common name	Notes	Swamp (S), lake (L), or riverine (R), fish brackish (B)	Stocked (St) or wild (W)
Anabas testudineus	Puyu, climbing perch	Favoured by farmers in Malaysia. Occurs in fair quantities in paddy fields, but seldom sold. Captural methods	S	W
Anguilla japonica	Eel	Occurs in Japan and India	SLR	St
Cambarus clarkii	Crayfish	Occurs in Japan	S	
Carassius auratus	Goldfish	Occurs in Japan		St
Catla catla	Catla	Indian major carp	R	
Channa striata	Snakehead	Common throughout South and S.E. Asia. Voracious predator, rarely stocked	S	W
Chanos chanos	Milkfish	Occurs in India and Indonesia	B	St
Cirrhina mrigala	Mrigal	Indian major carp	R	St
Clarias batrachus *Clarias macrocephalus*	Catfish	Occurs in Malaysia and other S.E. Asian countries	S	W
Cyprinus carpio	Common carp	Widely stocked as single species or in a polyculture. Occurs in China, Hungary, Italy, Japan, Madagascar, Pakistan, Spain and S.E. Asia	R	St
Helostoma temmincki	Kissing gourami	Commonly stocked in polyculture. Adults may be introduced to spawn in the paddy fields		
Labeo rohita	Rohu	Indian major carp	R	St
Lates calcarifer	Bhekti	Used in India		W
Misgurnus anguillicaudatus	U u	Philippines		St
Mugil corsula *Mugil parsia* *Mugil tarde*	Grey mullets	India and some isolated islands in Asia	B	
Mystus gulio	Tengra	India		

Oreochromis mossambicus	Tilapia	Stocked in Philippines in polyculture with common carp. All male stocking has been used	L	St
Oreochromis niloticus	Tilapia	As above. Mixed sex stocking. Stocking at 15 g to exceed 50 g in 100 days or at 20–25 kg in 80 days	L	St
Osphronemus goramy	Giant gourami	Herbivorous species, often used in polyculture in Indonesia. Introduced into rice fields as eggs, harvested at 8–10 cm. (Indonesia)		
Osteochilus hasseltii	Nilem carp			
Puntius goniotus	Tawes	Herbivorous species	R	St/W
Procambarus acutus acutus	White river crawfish			
Procambarus clarkii	Red swamp crayfish	Often considered a pest in rice fields in the USA, but has been farmed in some countries	S	W/St
Trichogaster pectoralis	Sepat Siam, snakeskin gourami	Often the most important economic species. Thailand and Malaysia		
Trichogaster trichopterus	Siamese gourami			

some attempts have been made to culture it. It is reported to be able to survive drought, long fallow periods and endure agricultural operations by remaining inside its burrow (Sommer, 1984). The rotational culture of rice and prawns in paddy fields that are seasonally brackish has been reported in Bangladesh. The prawns, mainly penaeids, although some species of the genus *Macrobrachium* are also raised, have become an important feature in the semi-saline zone. The polders are inundated with saline water during the dry winter months (5 months) when rice is not cultivated (Anon, 1985[b]). Fish culture may also productively use disused rice paddies. Paddy fields in which the water has become brackish are often used in this way for growing the snakeskin gourami (*Trichogaster pectoralis*) in Thailand. Brood fish are stocked at approximately 500 pairs/ha. For the first month of the production cycle, grass is encouraged to grow in the central parts of the ponds. Swathes are then cut in the grass and the cut grass heaped into rows. Subsequently, the grass areas are submerged to a depth of approximately 20-30 cm, so that the broodstock are stimulated to spawn. Water is exchanged several times during the production cycle and when the fish reach a size of 100 g (usually after 6-7 months) the fields are drained and the fish are concentrated on the peripheral channels and harvested by seine or dip nets (Lawson, 1981).

6.4 Rearing practices

In capture type rearing of fish, wild fish and fry will enter the paddy fields during the initial flooding and additional water exchange. Traps and screening devices may allow some form of management of the fish once in the paddy field. Where rice culture occurs in river valley or delta areas, this fish harvest can be considered as part of a flood-plain fishery. Its success therefore depends on adequate natural spawning, much of which must occur in the seasonally flooded rice fields. Thus, even though adaptation of the paddy field can allow concurrent culture of adult wild fish entering the paddy, or fish deliberately stocked, when rice production is intensified and modern management practices employed, natural spawning and fry production may be inhibited. The decline and reduced yields of the predominantly capture rice/fish systems in Malaysia and Japan support this (Tan *et al.*, 1973).

Both monoculture and polyculture are practised in rice fields, typical stocking ratios used in polycultures are given in Table 6.6.

Table 6.6. Typical polycultures used for fish-in-rice. (Source: Khoo and Tan, 1980.)

			Reference
Channa striata			
Trichogaster pectoralis		Malaysia	Tan and Khoo, 1980
Clarias batrachus			
Anabas testudineus			
Puntius goniotus			
Helostoma temmincki	35%		
Osteochilus hasseltii	15%	Indonesia	Ardiwinata, 1957
Cyprinus carpio	15%		
Puntius goniotus	35%		
Oreochromis mossambicus	50%		
Helostoma temmincki	20%		
Osteochilus hasseltii	10%	Indonesia	Ardiwinata, 1957
Puntius goniotus	10%		
Cyprinus carpio	10%		

The timetable of fish/rice culture is variable, depending on type of fish required and method (concurrent/rotational, see Figs. 6.5 and 6.6). In general, stocking of fry occurs around ten days after transplantation of rice from the nursery to the field, and fingerlings after three weeks (Singh *et al.*, 1980). Where rice culture is rain-fed, however, fish are sometimes stocked before transplantation to take advantage of areas of pooled water (Fig. 6.5) in the paddy field.

Direct seeding or the use of dapog seedlings necessitates an even greater delay before stocking with fish. The rice variety also affects the timing of fish introduction. Traditional varieties are usually tougher stemmed and less sensitive to the fishes' activities; time to maturation is also longer allowing a longer period of growth for the fish (around 160 days). Since these varieties are also taller than modern hybrids, a greater depth of water is generally kept in the field, which also favours fish production. More productive, high yielding varieties tend to mature earlier (105-125 days) affecting fish culture methods and the timing of stocking and harvesting.

6.5 Yields

Fish yields in rice fields, whilst not approaching those possible in a well managed pond of good depth (1-2 m), can be substantial; the total combined yield of fish and rice is particularly significant. Table 6.7 gives details of typical yields.

Production of fish is generally lower in captural systems than in cultural systems. This is mainly due to the slower growth rate of wild, principally swamp fish and sub-optimal stocking densities typical of captural fish-in-rice. Yields in fish/rice systems are highest when supplemental feed is given, the fertility of the soil and water is high (Table 6.8), the culture period is adequate and a polyculture of fish is used. In general, rotational rice/fish culture, when fish are stocked in fields without rice and can be more intensively cultured, yield more than concurrent culture.

The shorter culture period may be an advantage if fingerlings, with their rapid attainment of market size, are being produced. The high demand for fingerlings in Indonesia has prompted successful adaptation of rice/fish culture to intensive methods. Production of both fry and additional food fish may be considerable (Table 6.9). Pesticides are applied between fish crops and after draining the fields for weeding. Spiller (1985) noted that fish seed production has grown despite increasing imports of pesticides.

The best captural systems have yielded an average of approximately 135 kg/ha (Hora and Pillay, 1962). Khoo and Tan (1980) found that in Malaysia, the natural fish production ranged from 220-450 kg/ha with neutral pH, whereas in acidic water the fish yields varied between 11-56 kg/ha.

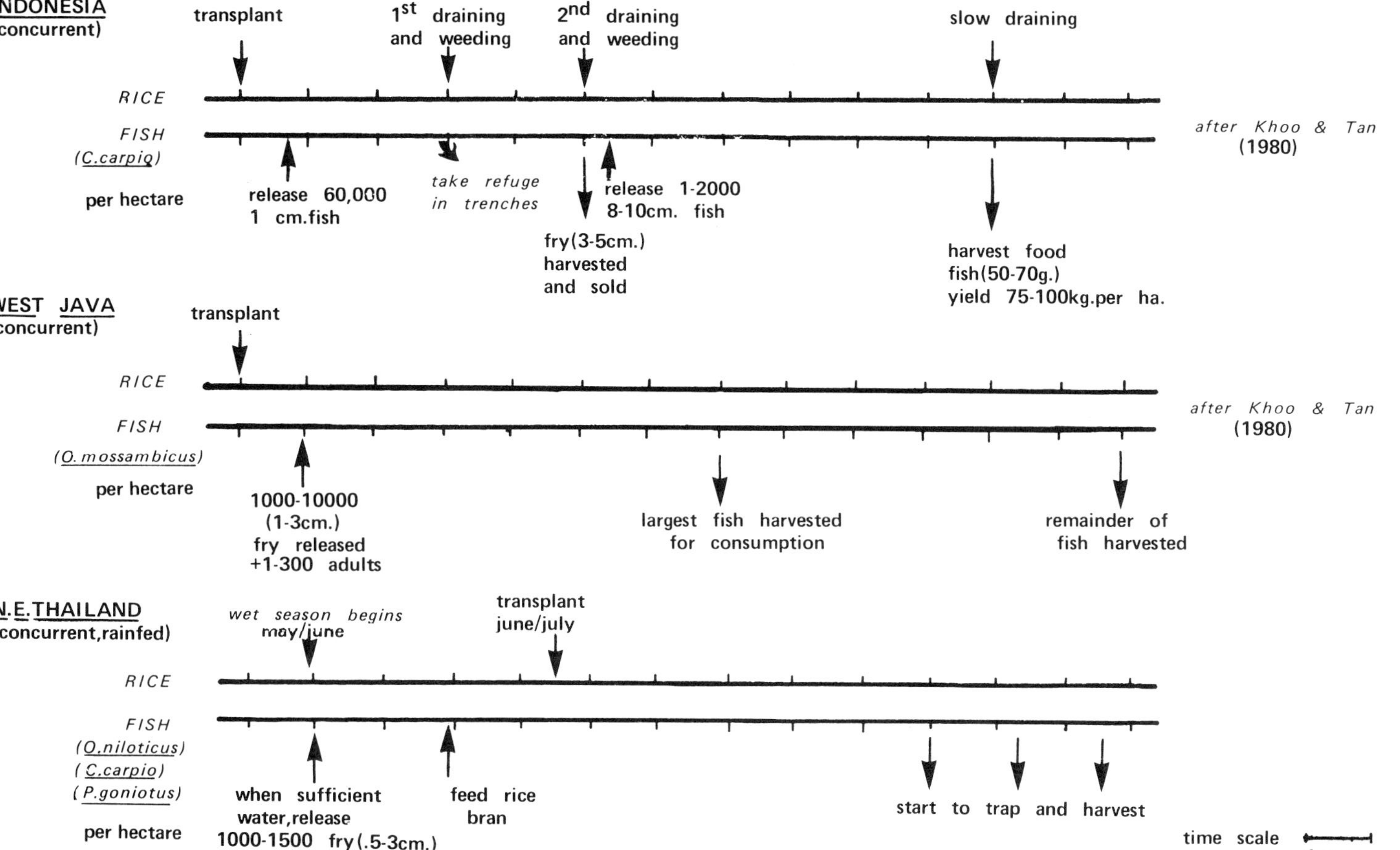

Fig. 6.5. Timetable of concurrent fish/rice culture.

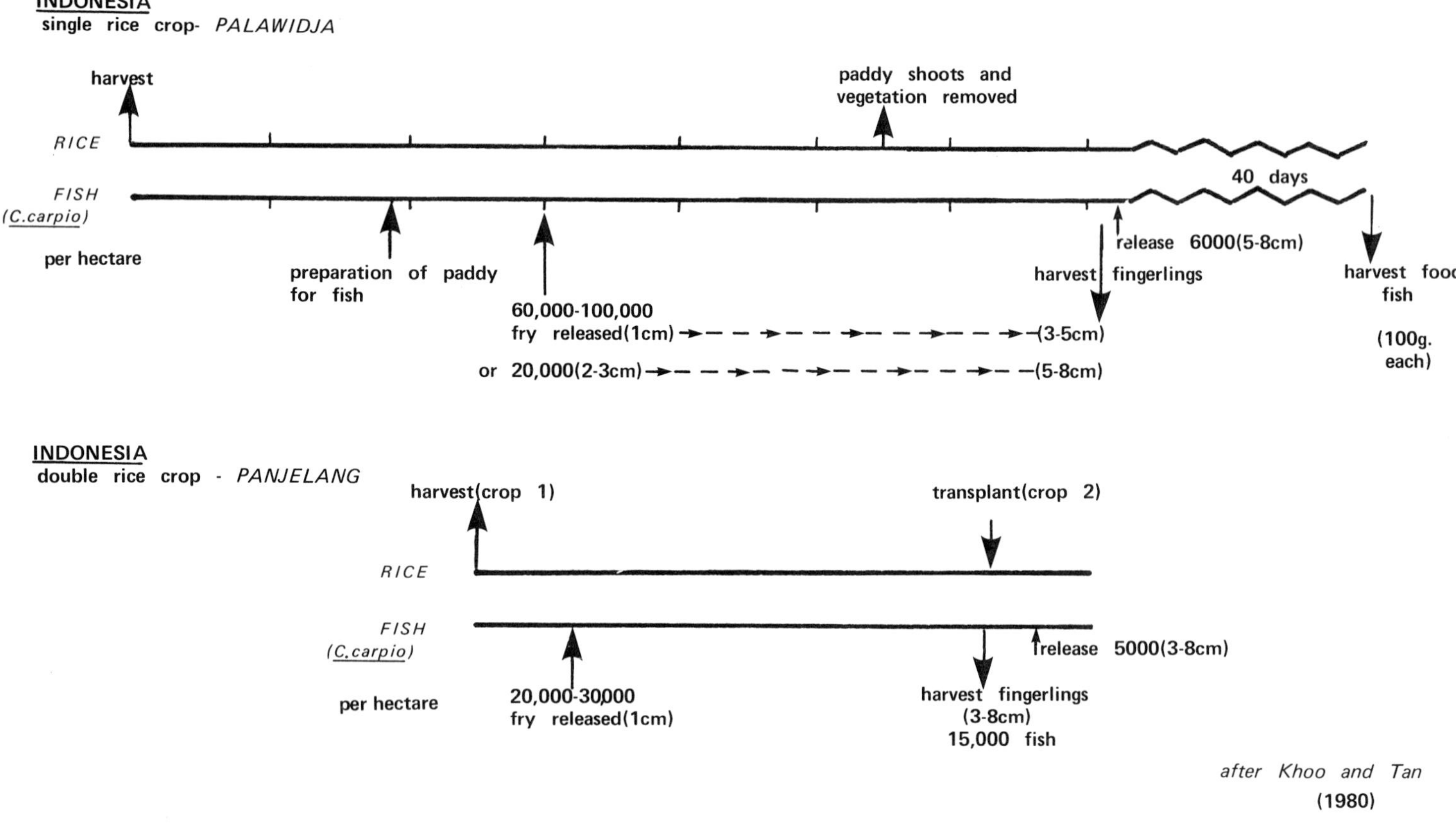

Fig. 6.6. Timetable of rotational fish/rice culture.

Table 6.7. Fish yields obtained from rice/fish culture systems.

Fish species	Stocking density (no./ha)	Fish yield (kg/ha)	Fish culture period (days)	Country	Source and notes
O. niloticus *C. carpio*	4000 2000	293	94	Philippines	De la Cruz, 1980 No supple-mentary feeding
P. goniotus *C. carpio*		60–100	75	Indonesia	Khoo and Tan, 1983
C. carpio	6000	600 – fertile water 300 – moderately fertile water 100-200 – infertile water	40	Indonesia	Khoo and Tan, 1983
C. carpio		220-450 – neutral pH 11-56 – acidic pH		Malaysia	Khoo and Tan, 1983

Table 6.8. The effect of water quality on fish yields from captural and cultural systems. (Source: Khoo and Tan, 1980.)

Captural	Yields (kg/ha)	Cultural	Notes
11–56			water acid pH (Malaysia)
220–450			water neutral pH (Malayasia)
		~ 100–200	infertile water (Palawidja, Indonesia)
		~ 300	moderately fertile water (Palawidja, Indonesia)
		~ 600	fertile water (Palawidja, Indonesia)

Table 6.9. Fry production using different methods in rice fields in Indonesia. (Source: Khoo and Tan, 1980.)

Species	No. of fry produced	Size (cm)	Additional food fish (kg)	Method
Cyprinus carpio	100 000–200 000	4–5	50–100	Palawidja
			40–60	Panjelang
	40 000–60 000	3–5	20–30	Panjelang
	20 000–30 000	5–8	20–30	Panjelang
Puntius goniotus	80 000–100 000	2–3	–	Panjelang
Osteochilus hasseltii	75 000–90 000	2–3	–	Panjelang
Helostoma temmincki	1 000 000	2–3	–	Panjelang

Battery chicken operation positioned over fish pond, Thailand.

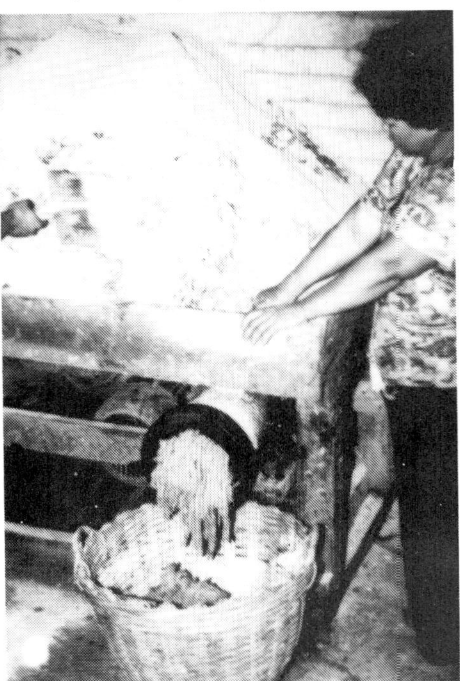
Grinding of marine trash fish for feeding to *Clarias* spp.

Aerobic composting at AIT.

Small-scale duck/fish pond in Pathun Thani, Central Thailand.

Housing and enclosure fabricated from bamboo. The ducks are the egg laying Khaki-Campbell variety.

Ducks gain access to swimming area via wooden walkway, in large-scale duck/fish farm, Central Thailand.

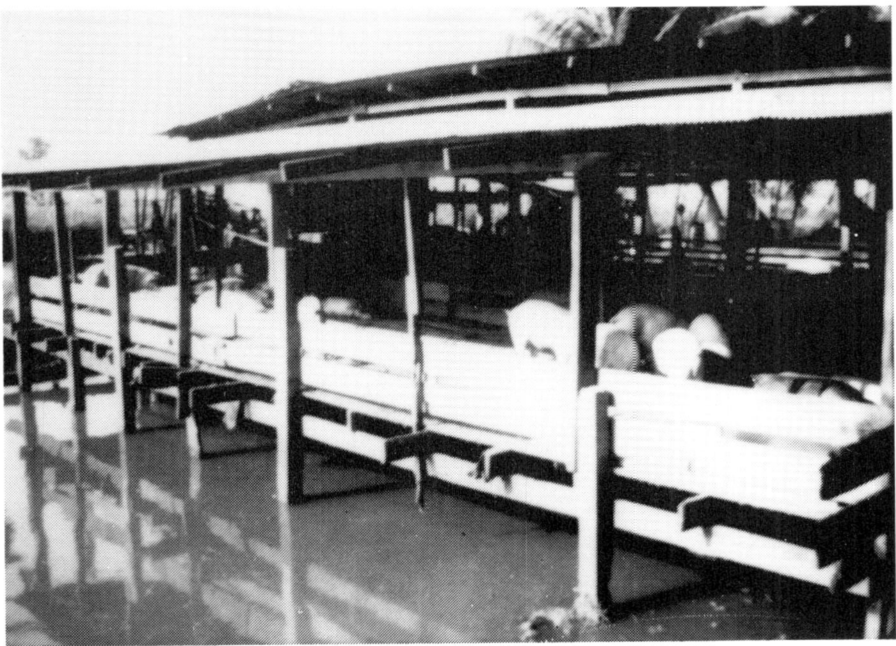

Pig fattening unit constructed on concrete supports over fish pond, Thailand.

Housing and enclosure fabricated from bamboo. The ducks are the egg laying Khaki-Campbell variety.

Positioning of latrine over fish pond.

Feeding of waste vegetables into fish pond.

Raised vegetable beds surrounded by irrigation channels for fish culture, outside Bangkok, Thailand.

Floating aquatic vegetation, mainly water hyacinth, cultured on the edge of canals and lakes.

Three-dimensional space utilisation by vegetable cropping on pond dykes in China (PRC).

Intensive vegetable (*Brassica* sp.) cropping on the steep banks of ponds in Southern China (PRC).

Utilisation of pond water for irrigation of vegetables on pond banks in China (PRC).

Biogas production using pig wastes on a state farm in China (PRC).

Enclosed feeding of macrophytes to Peking ducks over fish pond.
(*All photographs courtesy of P. Edwards.*)

CHAPTER 7
IRRIGATION/WATER RESOURCES

7.1 General considerations

Water is a critical resource for all parts of the agricultural system but the typically high demand for water by traditional fish culture may conflict with those of other users (Fig. 7.1). However, aquaculture can improve the efficiency of water use within the farm and even improve its value for integrated crop production. Conventional pond culture of fish may not be an option for some farmers on the grounds of lack of suitable land, cost of construction etc. The use of irrigation structures for fish culture has much potential and is reviewed below. Also discussed are some other aspects of fish culture integrated within the irrigated farming system. Weed growth can seriously affect water storage and distribution; the use of herbivorous fish as control agents has been proven technically and economically and could be implemented more widely. Fish ponds should not be considered 'black holes' from which fish is the only product. Wastes from fish culture, especially the nutrient rich water and sediments, can be conveniently used for on-farm crop production. The use and reuse of water between fish production and other activities can have negative effects and these are considered. The residues of pesticide use on crops returning to affect fish production via misapplication and run-off is discussed.

7.2 Fish culture in irrigation structures

Irrigation projects are usually planned primarily to benefit agriculture by improved water availability; often the methods for storage and distribution of water are under-utilised for fish production. This should be a key area of integration with fish culture in the future since, in addition to natural water, man-made resources are set to increase by a factor of six (to 12×10^6 ha) by the end of the century (Beveridge and Muir, 1987). Also, since the cost of conventional pond construction is often a limiting factor in the increase in fish production, more advantage should be made of water resources already available. When close to human settlement water resources frequently become eutrophic and thus their potential for low cost aquaculture is high. In arid areas where water is particularly valuable, competition amongst agricultural, domestic and industrial users may stimulate integration of water use. This can

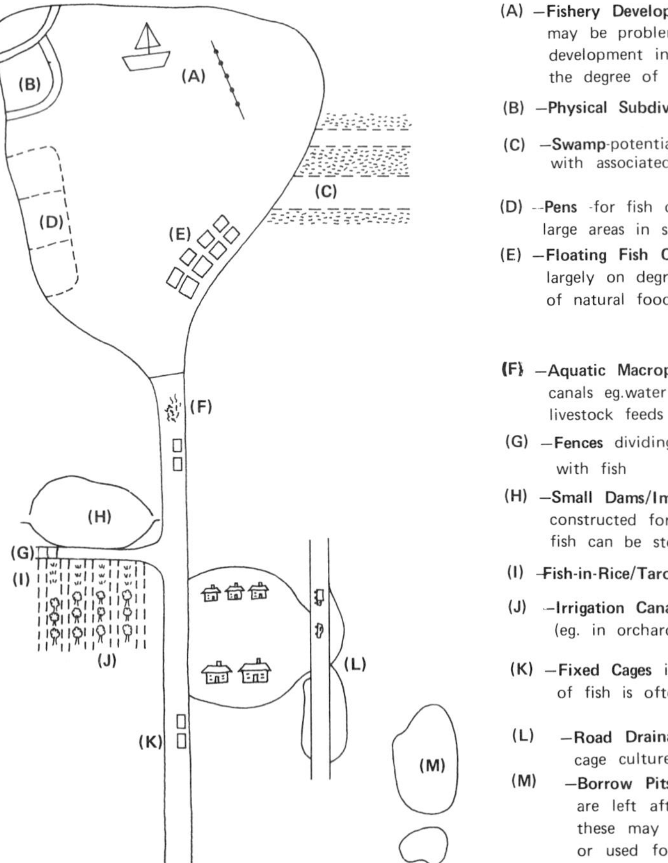

(A)	—**Fishery Development** in large eutrophic lakes and reservoirs may be problematic and unproductive. Urban and agricultural development in the surrounding area typically increases the degree of eutrophication.	
(B)	—**Physical Subdivision** into fish ponds	*People's Republic of China*
(C)	—**Swamp**-potential for *chinampa* development with associated aquaculture	*Mexico,Bangladesh, Burma,Kashmir.*
(D)	—**Pens** -for fish culture in shallow water;often enclose large areas in suitable lakes and reservoirs	*Philippines,Indonesia People's Republic of China*
(E)	—**Floating Fish Cages** -variable fish production,depending largely on degree of supplemental feeding and amount of natural food;site selection is often critical.	*Philippines,People's Republic of China*
(F)	—**Aquatic Macrophytes** can be cultivated in irrigation canals eg.water hyacinth,water lettuce;used for fish and livestock feeds	*People's Republic of China*
(G)	—**Fences** dividing canals ; partitioned areas can be stocked with fish	*People's Republic of China*
(H)	—**Small Dams/Impoundments**(5ha) on the farm; constructed for small -scale irrigation or livestock watering, fish can be stocked as in a conventional pond or in cages	*Isreal,Taiwan, Thailand,Zimbabwe*
(I)	—**Fish-in-Rice/Taro/Water Spinach** etc.	*S.E.Asia*
(J)	—**Irrigation Canals** within farms may be stocked with fish; (eg. in orchards)	*S.E.Asia*
(K)	—**Fixed Cages** in rivers,canals,sewers;-a high stocking density of fish is often possible	*Cambodia,Indonesia*
(L)	—**Road Drainage Channels** have potential as sites for cage culture;may be seasonal	*Panama*
(M)	—**Borrow Pits/Mining Pools**,often deep,ponded areas are left after extraction of minerals and aggregates; these may be managed as conventional fish ponds or used for cage culture	*Malaysia*

Fig. 7.1. A summary of integration of fish production with natural and man-made water resources.

be seen in Israel where more than 20% of the total fish culture area has been 'restructured' for water storage/irrigation purposes (Sarig, 1984).

7.2.1 Lakes and reservoirs

Although large amounts of water are often stored in lakes and reservoirs its use may be problematic. The scale of larger bodies of water ostensibly defies a conventional approach to fish culture; usually the resource is managed as a fishery.

On the basis of similar natural productivities, however, fish yields from lakes and reservoir fisheries are approximately twenty times lower than that possible in managed fish culture ponds (Beveridge, 1984). This is basically because pond management i.e. optimal stocking of fish feeding low in the food chain, elimination of predators and timely harvest of the crop is not possible on such a large scale.

Cage culture may provide a rational means of exploiting large public waters for the benefit of land-poor communities and the techniques can make intensive management and high yields possible. In the case of new reservoirs, there is often an urgent need to provide a livelihood for displaced lower valley dwellers. Cage farming of fish, as with pond culture, can vary in intensiveness (see Coche, 1982). The use of concentrated feeds and high stocking densities is becoming more widespread for species such as the Nile tilapia in the tropics, and the more carnivorous salmonids in temperate regions. Cages in eutrophic waters growing fish feeding low in the food chain have perhaps the greatest potential for low cost fish production. The more specific socio-economic advantages are discussed later (see section 11.5.2). Yields can be very high, since the productivity of the larger body of water can be tapped. In areas where natural currents and eddies increase the flow of water, and hence oxygen and food through the cage, yields of 1·9 kg/m^3/month during the best growing seasons are possible, e.g. Lake Buhi, Philippines (M. Beveridge, personal communication). This is approaching an upper limit for cage production, even if complete diets are fed. Average extensive tilapia production is in the range 1-5 kg/m^3/yr in the Philippines (see Table 7.1).

Where natural productivity and cage siting is poorer, supplemental feeding of low cost feeds may boost production. If natural productivities are low the cost of cage construction may be a limiting factor. There may also be conflicting requirements between cage operators and irrigation agriculturalists downstream.

Pens are another type of enclosure that can be used in shallow lakes. Used widely in the Philippines and China, they may combine cost advantages with those of intensive management. They may also allow more efficient utilisation of the natural feed resources since bottom feeding fish may also be stocked in a polyculture. The use of pens is believed to originate in Checkiang, China, where their development has since spread to almost 30% of potential sites (Tapiador *et al.*, 1977). Lake water pens are also well developed in the Philippines, principally for milkfish, but now for tilapia species. Collectives in the

Table 7.1. A summary of extensive and semi-intensive cage culture production of tilapia. (After Coche, 1982.)

	Species and situation	Stocking data (range)			Harvest data (range)			Culture period months	Remarks	Original references
		Mean weight (g)	(No. of fish/m³)	Biomass (kg/m³)	Mean weight (g)	Biomass (kg/m³)	Production (kg/m³/month)			
EXTENSIVE CULTURE IN EUTROPHIC WATER	*Oreochromis mossambicus* floating cages in Philippine lakes (commercial)	5–30	16–75	0·08–0·7	50–200	1·5–7·5	0·06–1·4	4–12	Production and culture period are largely determined by density of cages, siting with regards current flow and degree of eutrophication	Garcia, 1979 Guerrero, 1980 Coche, 1982
	Oreochromis aureus small (0·25 m³) cages in fertilised ponds (experimental)	3–25	500–600	1·7–12·5	27·3–73·6	17·9–54·8	6·9–18·1	2–3	Fish in 'moderate' plankton density showed low production and growth compared with those caged in high plankton density	Armbrester, 1972 Suwanasart, 1972
SEMI-INTENSIVE CULTURE USING PLANT MATERIALS AS SUPPLEMENTARY FEEDS	*Oreochromis niloticus* small (1 m³) cages in Indonesia	96–276	5 15 45	0·8–6·	170–218	1·1–7·6	(0·5) – (1·5)	3	Feeding *Hydrilla verticillata, Chara sp.* and *Lemna minor* at or above 30% bodyweight per day	Rifai, 1979, 1980
	Tilapia rendalli in Colombia	22·5	100	2·25	165	15·8	2·7	5	Feeding *Alocasia macrorhiza, ad libitum,* in 1 m³ cages	McLarney, 1978
		33·0	200	6·6	82	16·0	3·48	3+	Feeding *Alocasia macrorhiza* 10–20% bodyweight per day	Popma, 1978

Philippines control pens of up to 1000 ha in size, which have lowered natural productivity of the lakes and hence yields per unit area, and caused conflict with other interests. The encroachment of pens to the periphery of lakesides may further cause problems to the system's productivity, by disturbing the spawning and feeding ecology of natural fish using the fringes of the lake (Beveridge, 1984).

In China, the production from large lakes and reservoirs has been intensified by other means. Subdivision by land fill and dredging of shallow lakes and reservoirs into manageable areas of fish production has been attempted in some areas (Tapiador *et al.*, 1977).

7.2.2 Canals and rivers

Culture of fish in canals and rivers is not a common practice despite their proximity to and physical integration with farming systems. Hickling (1962) has described cage culture of *Pangasius* spp. in rivers in Cambodia and this also occurs in Thailand. Supplemental feed such as trash fish, bananas and rice flour is given, and, because of the free exchange of water, dense stocking of fish is possible (80 kg/m^3).

The culture of common carp in fixed cages placed in stream and canal beds enriched by domestic and agro-industrial wastes has been reported in Indonesia (Vaas and Sachlan, 1956). Yields, with minimum feeding, can exceed 50-75 kg/m^3 in 2-3 months. Coche (1982) reports that the practice has also been tried on a smaller scale with the Mossambique tilapia (*O. mossambicus*) but that yields (at 0·55 kg/m^3/month) were only about half those possible in eutrophic lakes (Table 7.1).

The placing of caged tilapia in irrigation channels has also been reported in arid regions such as the Middle East (Hopkins, 1983).

Problems associated with this type of integration often stem from the demands of irrigation. Canals may be designed, and rivers altered, for fast flows unsuitable for cages. Alternatively, flows and capacity may be irregular in accordance with the needs of irrigation. Cages may also cause adverse sedimentation or flooding in canals. Larger channels have been divided by fences in China (Tapiador *et al.*, 1977) and fish cultured extensively between them. Yields are reported to be as high as 1500 kg/ha (300-1500) with regular stocking and feeding of supplementary grasses and manures. On farms, however, irrigation supply channels are not used for fish culture because of the constraints suggested above. These canals and ditches are used for intensive culture of aquatic macrophytes especially duckweeds and water hyacinth. Simple release of *Oreochromis* spp., *Channa striata* and *Puntius goniotus* was practised in the irrigation channels of around 450 orchard and vegetable farms in Central Thailand; resulting in an estimated 350 t/yr (Swingle, 1972). *Macrobrachium rosenbergii* was also monocultured in orchard irrigation canals in the same country. The culture of bighead carp, grass carp and Nile tilapia at low density in the irrigation channels supplying vegetable beds has given variable results in Thailand. Edwards *et al.* (1986) reported high mortalities of the carps (40%), although the survivors grew well to a mean weight of 1 kg/fish in seven months. Possible reasons for the poor survival include lack of feed, heat stress due to the relatively shallow ditches, or the effect of toxic pesticides used on the vegetables.

7.2.3 Other man-made resources

Cage culture, by its versatility, is well placed as a method to exploit seasonal surplus of water, such as occur in waste water and roadside drainage canals. Of the estimated 5000 ha of roadside canals in Thailand alone, most remain at least seasonally inundated. Cage culture may also have particular value in allowing integrated use of fish culture in disused mining pools. These tend to be deep, undrainable and often eutrophic. Tan *et al.* (1973) reported that 78% of the area devoted to fish culture in Selangor, Malaysia, is in the form of disused mining pools; after the release of the fish into the mining pools minimal management is practised. Most mining pools were unimproved but still showed a better return than excavated fish ponds because of the low capital investment.

The potential of extensive aquaculture on river flood plains in South America and Africa has also received attention. Welcomme (1979) reports that dyke systems, called 'modulos' already conserve water through the dry season over some 30000 ha of flood plains in the Orinoco and Amazon area. The gradual drying off of the plain maintains areas of fresh pasture for cattle ranching and could allow culture of fish within the ponded areas.

Smaller dams, weirs and ponds, primarily designed for irrigation purposes or livestock watering, may also be important centres of fish production and can be managed in the manner of an intensive fish pond. Fish culture is practised thus in livestock water supply ponds in Zimbabwe and irrigation ponds in Israel and Taiwan. Chen and Li (1980) reported that about 2000 ha of irrigation ponds in Northern Taiwan alone were used in this way for polyculture of Chinese carp. A large turnover in water prevents a build-up in the primary production although fertiliser application (superphosphate) can substantially improve the yield of phytoplankton-feeding fish. In Sri Lanka and India, significant fish yields are obtained in seasonal 'tanks' or reservoirs, usually stocked as a polyculture. Yields are particularly improved by the growing of fodder crops and the grazing of cattle and other animals on the pond bed during the dry season. In Israel, irrigation has followed on from the development of fish culture, and so conventional fish ponds have been enlarged to serve as irrigation structures, whilst continuing to function as fish production units. Growth and production of fish is reported to be greater in such ponds or reservoirs than in conventional ponds, using normal Israeli polyculture and feeding/fertilising regimes. Undoubtedly, the greater volume and depth (3-7 m) serve to reduce the concentration of growth inhibiting metabolites that occur in conventional intensive Israeli pond systems. Management strategies are aimed at accurate initial stocking of fish (Table 7.2) and 'thinning' as the water is gradually used up for irrigation. A backup of nursery ponds (for small fish) and holding ponds (for market fish) give the system flexibility for the needs of both fish and crop grower (Sarig, 1984). Future designs of integrated systems may further improve the production potential of irrigation dams and tanks (see section 12.5.1).

Table 7.2. Stocking densities and yields of polycultures in Israeli irrigation reservoirs for different culture periods. (Source: Sarig, 1984.)

Culture period (days)	Stocking density (fish/ha)					Yield (t/ha)
	common carp	tilapia	mullet	silver carp	total	
250–300	4000	13250	2000	500	19750	12·91
250–300	3600	2500	1800	550	8450	12·65
250–300	6900	10000	2000	500	19400	11·24
150–200	4300	1000	2000	300	7600	6·68
150–200	6900	–	2000	200	9100	6·16
150–200	3980	1000	1990	300	7270	5·31

7.2.4 Swamps

Swamps and wet lands are often considered to be wastelands. However, detailed analysis often shows them to be extremely productive ecosystems in their own right, and potentially capable of providing large nutrient outputs into surrounding waters. They may also be important in the breeding ecology of bordering lake and reservoir fisheries. Traditional methods to tap this fertility in areas where arable land is lacking include the Chinampa system, as used by the Aztecs (Ruskin and Shipley, 1976). Floating vegetable gardens are produced by making mats of aquatic vegetation covered with bottom nutrient rich mud. Thus, a productive natural system is harnessed for intensive food production without the swamp's destruction. The Chinampas are variable in size and in the land/water ratio used. Pullin (1982) reported efforts to integrate fish culture with them in the state of Vera Cruz, Mexico. The fertile water channels between the raised strips would be stocked with herbivorous tilapias, and managed for their production. Since such floating gardens are also used in Bangladesh, Burma, Kashmir and parts of the Philippines (Ruskin and Shipley, 1976; Ramsey, 1983), the method may have considerable potential for integrated food production.

7.3 Control of weed macrophytes in irrigation systems

Much interest has focused on the use of herbivorous fish for controlling weed infestation, particularly of irrigation channels and reservoirs (Table 7.3). Infestation in irrigation systems may quite dramatically affect the productivity of the larger agricultural system they are designed to improve. In one irrigation scheme in Asia, for instance, submerged vegetation reduced water flow in the main canal by 80% within five years (Holm *et al.*, 1969). Feeder canals to individual fields, with lower flow rates, are liable to suffer even greater reductions, and extensive culture of fish within these may considerably improve the situation. Considerable improvements in weed control using the grass carp have been claimed in Egypt (Huisman, 1983). In addition to an annual harvest of 200-350 kg/ha of grass carp, improved production of endemic fish species (extra

Table 7.3. Control of aquatic weeds by herbivorous fish. (Adapted from Edwards, 1980[c].)

Species	Introduction/purpose	Result	Source
Ctenopharyngodon idella	Malacca – to clear 1·8 ha pond 375 fish stocked	Estimated 22 t cleared in 110 days	Hickling, 1960
	Russia – to clear Kara Kum Canal	Planned flow reduced by weeds, an estimated loss of 20 000 ha of irrigated cotton. Notable decrease in weeds after stocking.	Edwards, 1980[c]
	Arkansas – to clear 20 000 ha public lakes	After 15 years, no problem infestations	Ruskin and Shipley, 1976
Tilapia rendalli and *T. zillii*	2-10 ha reservoirs Kenya	Total eradication after 2·5-3 years	Van der Lingen, 1968
T. zillii	Canals of Imperial Valley Southern California 2500 fish/ha stocked	Complete elimination	Ruskin and Shipley, 1976
Tawes *P. goniotus*	E. Java, Indonesia irrigation dams	Cleared 284 ha reservoir covered with *Ceratophyllum* and *Najas* in 8 months	Schuster, 1952
Giant gourami *Osphronemus goramy*	India irrigation wells	Control of submerged weeds	Hora and Pillay, 1962
Tilapias	In Hawaian sugar cane plantation irrigation canals 75 000 7·5-10 cm fry	After previously using herbicides (annual cost $5000) – reduced to stocking cost of fish only ($3000). Prevented regrowth of weeds	

100-150 kg/ha/yr) and a reduced incidence of the snails transmitting bilharzia were achieved. Furthermore, Huisman claims weed control by grass carp is economically favourable (section 11.5.3).

Stocking rates

Stocking rates for weed control should ideally maintain a balance between consumption and growth, so there is little obstruction to flow but enough food for the fish. Also, if the fodder is to be relished by the fish an attempt to maintain the plants at their most succulent, normally the exponential (active growth) phase of their growth cycle, should be made. Since aquatic macrophytes are often present in varying combinations and amounts, and the fishes' preference changes, this balance may be difficult to achieve in practice. Controlling the stocking rate and size of herbivorous fish is a frequently used method. Mehta *et al.* (1976) suggested that 100 grass carp (stocked at 1·0-1·5 kg body weight) per hectare was suitable. Larger numbers of the smaller, less voracious *Puntius* species or tilapias would be required for the same effect. Huisman (1983) records the stocking of small grass carp individuals (30 g) where predatory species of fish are few, and larger specimens (100 g) when waterways are drained annually. Small grass carp (from 1·5 g) have been used successfully for weed control in drainage canals in Egypt (Van Weerd, 1985). Regrowth of submerged weeds was completely controlled at stocking rates of 100 kg/ha of infested waterway; small fish are more efficient weed controllers than larger grass carp since the same quantity of fish have more mouths that consume the growing points of plants (Zonneveld and Van Zon, 1985). When the presence of predators restrain the stocking of small fish these authors recommend the use of 250 kg/ha of large (over 200 g) grass carp.

Unfortunately, some of the weeds least favoured as food by the fish are also some of the most common and troublesome e.g. water hyacinth (*Eichhornia crassipes*). Some authors have reported that grass carp will eat it only when it is the only weed present (Avault *et al.*, 1968). The most effective strategy is thus to stock herbivorous fish as the macrophytes first appear, or, if already dominant, remove by mechanical or chemical means (see Ruskin and Shipley, 1976) and then prevent regrowth by immediately stocking herbivorous fish.

7.4 Use of fish culture waste water and silts

The integrated pond system is often considered as an efficient means of recycling otherwise wasted nutrients. Nutrient analysis, however, shows that their recovery is far from complete, and that accumulation of these valuable substances occurs within the pond (Little, 1983) (Fig. 7.2). A study of the typical pond environment (section 1) reveals that only low concentrations of dissolved inorganic nutrients (in particular nitrogenous and phosphate compounds), remain available in the water. Most of these compounds are quickly and readily absorbed into the food web before eventually accumulating in the pond sediments.

Intensive tank culture of fish is an increasingly common practice even where extensive forms of culture are still used. Although high quality, complete feeds are commonly used, typical food conversion ratios show that large amounts of residues are produced as uneaten feed and fish waste. Alterations of the ecology of receiving waters, and also nutrient loss from the system are inevitable consequences. Efficient separation and collection of solids can result in some nutrient recovery, and a reduction in pollution. Loss of soluble excretory and leached products still occurs, however.

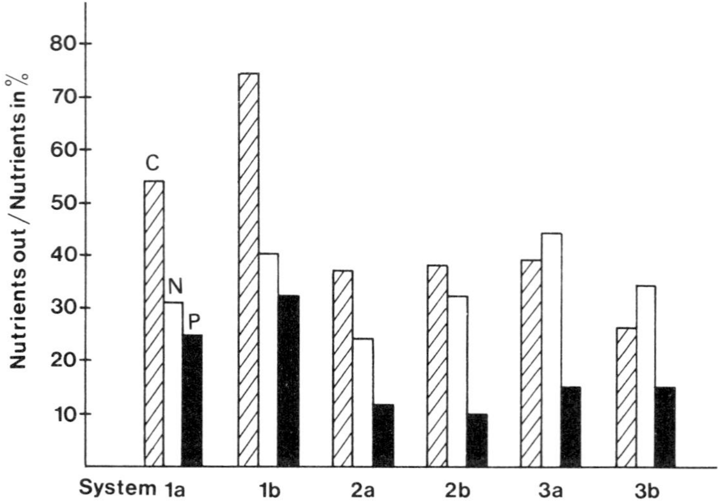

Results of a theoretical study of nutrient recovery in three integrated farm systems in Thailand.

System 1 Simple village level system
- a) Wooden pens/pigs
- b) Rice/pig/fish

System 2 Intermediate scale system
- a) Rice/fish/pigs, 1 ha fish pond
- b) Rice/sorghum/maize, guinea grass/pigs/ 5 ha ponds

System 3 Large-scale systems
- a) Components of form closely integrated
- b) Components of form physically separated

Fig. 7.2. Nutrient recovery in integrated pig/fish systems. (From Little, 1983.)

7.4.1 Pond culture effluent

Little attention has been paid to the potential of fish pond sediments and water for use in associated agronomy. The major exception is China where a shortage of productive land, water and fertiliser have forced the trend over decades, if not centuries. The ability of ponds to reserve nutrients makes the fish pond a fertility sink, which can usefully support adjacent land cropping. It has been estimated that river and pond muds furnish 6·7% of total nitrogen, 7·3% of total phosphorus and 4·4% of the total potassium used in the whole of the country's agriculture (Plucknett and Beemer, 1981). In areas where fish ponds are more densely concentrated, this source will be considerably more important. Factors that are especially condusive to China's efficient use of pond residues are listed (see Table 7.4). Analysis of these factors perhaps explains why the practice is not widespread, even though the most important aspects (1-4) are reproducible or are in current practice in other countries.

Table 7.4. Summary of factors encouraging use of fish pond sediments and water for agriculture in China.

(1) High cost/poor availability of chemical fertilisers for agriculture.
Chemical fertilisers are used increasingly, but as a proportion of total inputs they are still small, e.g. 24% total nitrogen:total phosphorus, and 1% potassium (Plucknett and Beemer, 1981). The most important sources of organic manure are animal and human wastes, green manure, compost and pond muds.

(2) Plentiful labour.
Removal of sediments and manual irrigation of crops is very labour intensive – China's high population is rurally based and has allowed this strategy.

(3) Fertile water and deep pond sediments.
Intensive use of organic inputs, particularly those of plant origin, can result in a rich and continuous sediment. Climate and management practices certainly affect sediment accumulation. The continual addition of wastes during periods of cool weather leads to their build up in the pond. In the tropics, where temperatures remain more constant, sediments accumulate more slowly despite high loadings (Hopkins and Cruz, 1982).

(4) High dyke:water area ratio.
Pond dykes up to 10 m in width have been reported as typical (Coche, 1980). Pond effluents are used in the immediate area of their production.

(5) High degree of fish pond management.
Ponds are regularly drained and maintained. Sediment removal thus becomes part of pond management, maintaining capacity and reducing organic and nutrient loading that can cause growth reduction in fish.

(6) Irrigation
Fish farms are normally irrigated. Thus both water quality and depth, can be maintained for fish culture, despite its use for irrigating bankside crops (Tapiador *et al.*, 1977).

(7) Electricity.
Intensive rural electrification allows for economical use of water and sludge pumps. The latter provide a method of sediment removal that is less labour intensive than traditional means.

(8) High levels of crop production and aquacultural skills on farm.
Specialisation of personnel at the commune level allows a technical approach and high yields of both fish and crops.

(9) Crop production; techniques and varieties used have evolved with fish culture.
Numerous authors have reported the diversity and yields of crops grown adjacent to ponds. Often grown as fodders for fish and livestock, the concept of 'rational close spacing' is often used; the crops are grown in patterns as intensive as is feasible. Transplanting, multiple cropping and intercropping techniques are used. Trellising to utilise three-dimensional space is common with climbing plants overhanging the pond (Edwards, 1982[a]). Cultivation of perennial fruit and mulberry next to ponds is also widespread as is sugar cane.

Pond water

The use of pond water for small-scale irrigation is practised in a number of countries. This pond water may be particularly valuable where rainfall is seasonal; in the subsistence situation this may allow continual cropping of household vegetables throughout the dry season. Fish pond water for its irrigation value may then be of greater importance than its potential contribution to fertility. At the kibbutz and farm level in Israel, for instance, every effort is made to re-use aquaculture water in crop production. The ponds primarily for irrigation are fertilised with sewage and stocked with fish, to supply dry season field crop irrigation in Israel (Hepher and Pruginin, 1981). Small-scale production of high value fruit, mulberry and jasmine, using pond waters is also common throughout South East Asia, and may provide significant cash benefits (Tan and Khoo, 1980). Aside from the considerable irrigation benefit, the content and value of the nutrients in the water contained largely in the phytoplankton should be considered.

Edwards *et al.* (1980) assessed the use of green fish pond water for cultivation of maize (*Zea mays*). Compared to maize fertilised with raw sewage or chemical fertilisers, yields were low since production was well correlated with nitrogen applied, and the pond effluent had a lower concentration compared to the other treatments (Fig. 7.3). The phytoplankton concentration, and thus nutrients in the effluent water, was relatively low because of initially low fertilisation of the experimental ponds. More intensive direct fertilisation of fish ponds would be likely to improve nutrient quantity and irrigated crop response. Using even the 'dilute' fish pond water over the 15 week growing period, a total average of nearly 40 kg/ha of nitrogen was applied to the maize. The use of this fertile pond water improved yields of maize to nearly double (3600 kg/ha/crop) that of unfertilised yields (2000 kg/ha/crop). Where such crops are usually grown with little or no fertiliser, this is a considerable gain. Furthermore, if irrigation is available from adjacent fish ponds, three crops a year may be gained compared to the single crop normally grown. Even where chemical fertilisers are used intensively, the addition or substitution of nutrients from pond water may at least reduce cost and dependence.

The use of fertile fish pond waters to irrigate rice was documented by Le Mare (1952), and yields comparable with chemically fertilised systems (urea) were claimed in the Philippines for rice and vegetables (Eusebio *et al.*, 1978).

Pond sediments

Strategies for using sediments for crop production may involve:

(a) removal of sediments from the ponds for use on land, in particular pond banks,
or
(b) draining of most, or all, of the water and growing of crops within the pond itself (Fig. 7.4).

(a) In China, sediments are removed whilst the pond is full with water and

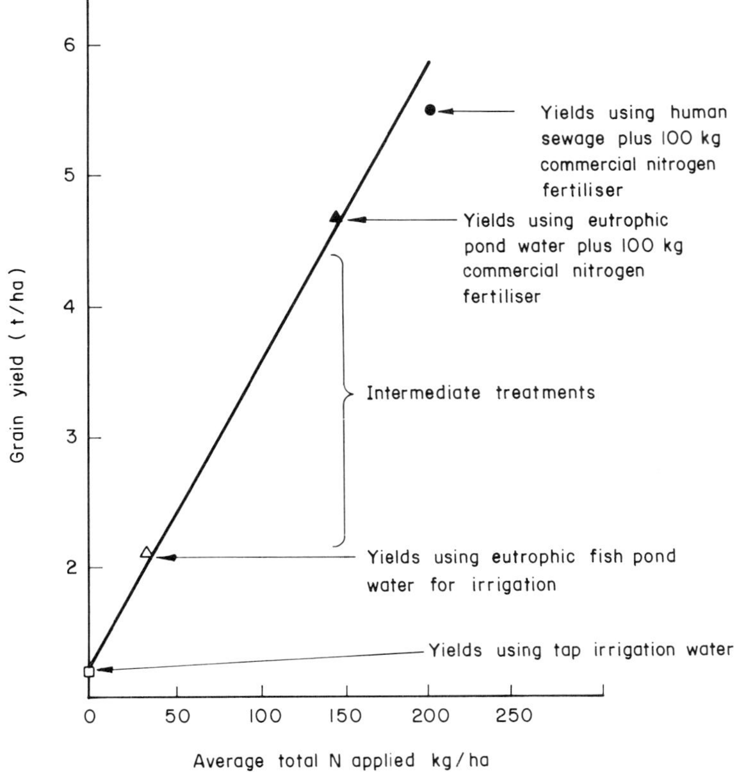

Fig. 7.3. Relationship between maize grain (*Zea mays*) yield and total applied nitrogen (after Edwards *et al.*, 1980).

fish are being cultured. This may cause problems with turbidity if practised too regularly; although as only part of the pond is cleared at one time some overall benefit may be gained by resuspension of settled nutrients. More normally sediments are removed by sludge pump, or, laboriously, by hand. The drying out of the pond to enable removal of sediments would be expected to excacerbate losses of volatile nutrients, especially nitrogen as ammonia. The main advantage of this method is that fertilisation of crops may be achieved during the normal routine maintenance of the pond, and thus fish production need not be affected.

(b) Draining may not be necessary, if a dry season and/or use of water for irrigation results in gradual and seasonal draw-down of water. Fish production ceases entirely during these times. Rotation of fish ponds for fish and arable cropping is a traditional European practice (Hickling, 1962) designed to maintain pond fertility and re-use nutrients. It may have considerable potential in drier areas of the tropics where only seasonal fish culture is possible. The residual moisture, as well as the nutrient value of the sediments, contributes markedly to the yields possible. Delmendo (1980) describes a rotation system

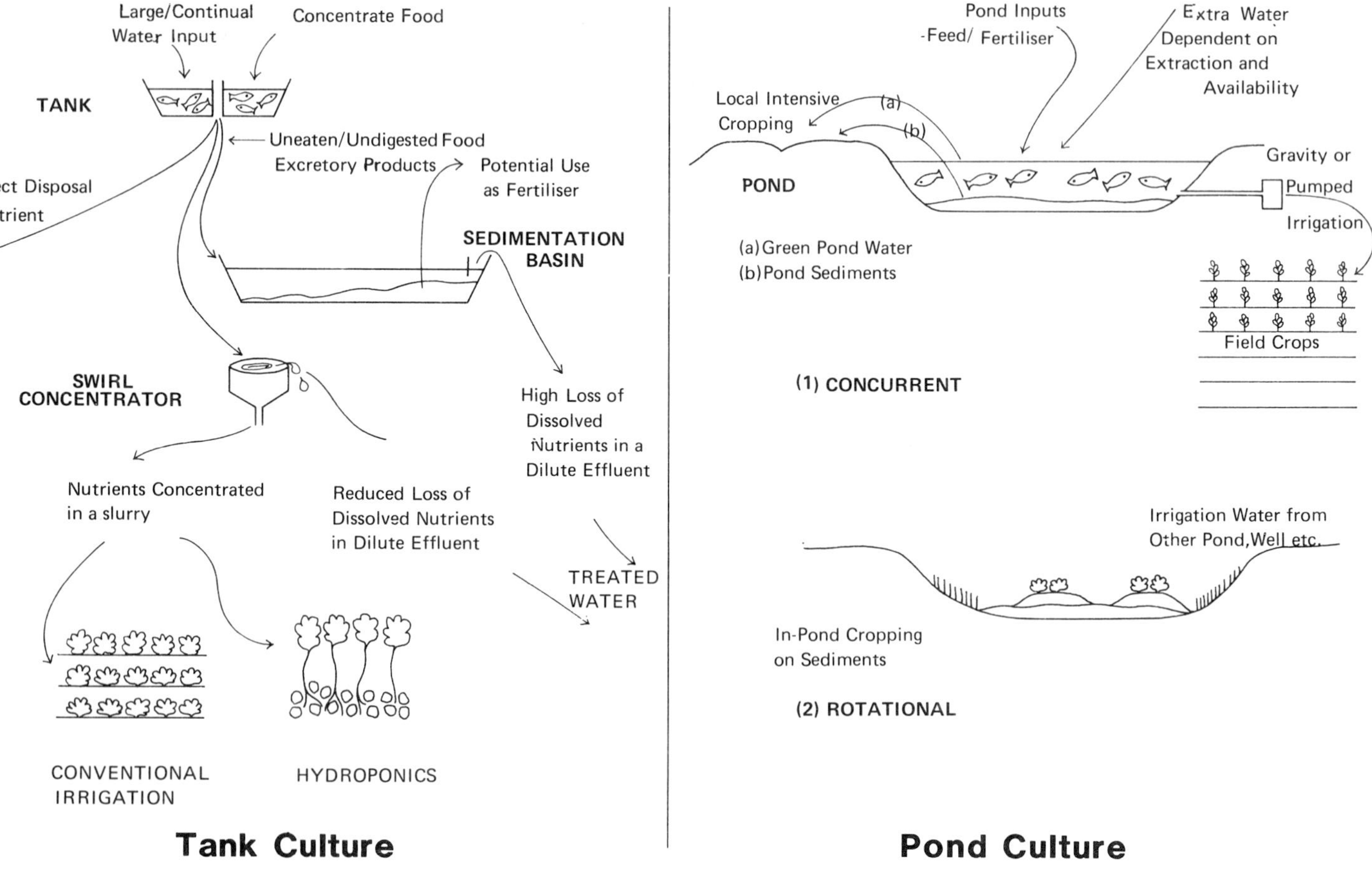

Fig. 7.4. Strategies for recovery of nutrients from pond culture of fish via plant cropping.

in Central Thailand, where the pond bottom of two ponds is alternatively cropped (section 12.3.3). However, Edwards *et al.* (1986), on practical testing of this system, found the rotation of ponds to be impractical; the preparation of the wet muddy pond sediments for vegetable production is highly labour intensive.

A much longer-term rotation system has been used in Hungary, although not adopted on any commercial scale. Duck/fish, fodder crop and rice production are alternated over a period of six years each. The major disadvantage, however, whenever water is available year-round, is that fish culture is usually more profitable than other agricultural uses. The capital cost of the pond construction is usually high; and the growing of relatively low value crops would not give the best return. Nonetheless, there are advantages to this method (section 12.3.3).

Efficiency of nutrient removal through plant crops may still be low since the sediments contain such an abundance of nutrients compared to the crop's potential rate of removal and the time over which they are grown.

Exposure to sun and wind will also reduce volatile nutrients substantially (Degani *et al.*, 1982), especially if the soil is ridged up and cultivated conventionally. This is necessary for such crops as maize, which are sensitive to excess moisture (Michael, 1978). Other crops, such as fodder grasses, beans (*Phaseolus* spp.) may be planted straight into the sediments with success.

Unfortunately, few data are available concerning yields, although Muller (1978) claims rice yields following legumes cropping and fish/duck culture were 50-100% higher than average production in Hungary. Hepher (1952) has also tried dried pond cropping with some success, both for grain (barley and maize), green manure (barley) and vegetables. Using minimum tillage, a satisfactory green manure of barley was obtained with 50 kg seed/ha; fish were introduced after two weeks with satisfactory results. Grain yields of 2500 kg seed/ha were taken. Over-watering caused losses of maize but where this was avoided 8-9 t/ha of grain and cobs resulted. Cauliflowers yielded 7-8 t/ha over a three month cropping season.

7.4.2 Tank culture effluent

Whereas pond sediments and water may be seen as a bonus resource to local agriculture, effluents from tank systems are more likely to be considered as pollutants. Tank effluents are also produced continually, and this may pose problems for their use. In general, tank culture of fish will involve large quantities of water and discharge of 'waste water' and thus will have greater interaction with natural waterways. If left untreated, nutrients in the form of uneaten and unassimilated food, together with excretory products of the fish, will be lost in the effluent water.

Methods to reduce these effluent wastes, and the undesirable BOD and nutrient loading they inflict on receiving waters, are largely the result of pressures to alleviate pollution. They basically rely on physical separation of waste solids from effluent waters and do not attempt removal of the dissolved fraction. Sedimentation basins are the normal means of treatment, resulting in a sludge of high BOD and nutrient quality. However, this is at best an ineffective method of nutrient removal, since further breakdown of particulate matter and release of soluble nutrients occurs in the sedimentation basin. Clarke (1983) proposed that if efficient self-cleaning culture tanks are used with a swirl concentrator much more efficient solids removal with less leaching can be expected. Moreover, since the solids are produced as a concentrated nutrient slurry, they can be more easily incorporated in conventional irrigation or hydroponics.

An approximate estimate of the available nitrogen and phosphorus has been made by Ketola (1982) for salmonid waste. Approximately 22·8 kg of phosphorus is produced per tonne of fish raised, using a standard feed (FCR 1·2). Of this about one third (7·85 kg) is available in a sludge using means of separation and collection. Using a calculated estimate of nitrogen concentration (using total nitrogen: total phosphorus) 55 kg of nitrogen is produced per tonne of fish raised. Thus, for a 100 t/yr unit, 5 t of nitrogen and nearly 0·8 t of phosphorus are available per year, using simple, presently available technology. The potential value to local agriculture of nutrients in this concentrated form can be appreciated if for instance maize yields are considered with respect

Box 7.1. Use of tank effluents for pond culture.

Typical intensive tank culture production,
e.g. with 10 t of fish being held, produces
~ 10 t × 0·025 (feeding rate 2·5% bodyweight/day)
× 0·04 (% total nitrogen in feed that is wasted)
× 0·002 (% total phosphorus in feed that is wasted)

Therefore, approximately 10 kg nitrogen/day and 0·15 kg phosphorus/day are present in waste waters from 10 t culture unit.

In nitrogen terms, this is equivalent to ~250 pigs and would support fish production in about 2·5 ha of ponds.

However, water flow in intensive systems is high;
in this case ~0·8 m^3/min; = 8 m^3/min for 10 t.
Thus the time which the water from the tanks would spend in the ponds would be;

in a pond of volume (25000 × 1·2); 30000 m^3, water enters at a rate of 8 m^3/min; time spent is 30000/8 or 3750 min i.e. 3750 (60 × 24) = 2·6 days

This residence time might not be sufficient for all the nutrients to be used properly, so there may be choices:

(1) accept reduced productivity, i.e. lower pond yield

(2) make bigger ponds and use a supplementary nutrient source to bring up the yields

(3) concentrate the wastes, e.g. by settling out effluent solids, and so reduce the water flow associated with the wastes

to total nitrogen applied (Fig. 7.3). On a site of approximately 10 ha, a yield of 40 t/crop might conservatively be expected (three crops a year).

Recovery of the soluble fraction using hydroponics has been found ineffective mainly because of their dilution. Some possible ways in which tank culture of fish might be incorporated into irrigated agriculture as an integrated component are suggested later (section 12.5).

7.5 Pesticide use and plant/fish integration

The widespread use of pesticides and their contamination of water courses is a cause of major concern to fish culturists. Persistent pesticides are believed to be the major cause of reduced yields of fish in rice fields in Malaysia and Indonesia (Khoo and Tan, 1980; Koesoemadinata, 1980), but the potential problems are not limited to this particular form of fish culture. Aquaculture within irrigation systems is at risk from upstream contamination, and ponds fed by rain and local run-off receive pesticides used nearby. Large scale endosulphan applications over 133000 ha in East Java as part of a rice improvement project were followed by fish kills in ponds and rivers downstream, causing considerable economic loss (Gorbach *et al.*, 1971). The problem may not only manifest itself in the acute toxic form however; chronic effects of reduced availability of natural foods, growth rate and fry survival are also possible. Koesoemadinata (1980) has established that amongst the three main groups of pesticides currently employed, carbamates are less toxic than organophosphates and organochlorides (Table 7.5).

Toxicity is assessed using standard LC50 tests. The majority of pesticides have a broad spectrum of action and adversely affect many organisms, including fish. The severity of the reaction depends on the rate of pesticides entering into the fishes' environment, i.e. the rate and frequency of application. A pesticide's stability to degradation also varies; many are highly persistent and their use can result in widespread and long-term contamination of the environment. Chlorinated hydrocarbons for instance have half-lives of 10-15 years. The majority of pesticides are reported to be insoluble in water but highly soluble in organic solvents; accumulation in biological tissues can thus occur (Table 7.6).

Methodology for pesticide use in fish culture

Pesticide use and fish culture are not totally incompatible, however, as has been shown by Estores *et al.* (1980). High run-off of pesticides from agricultural land is symptomatic of ineffective application and use of the pesticide. The use of pesticides in granular form may be preferable to foliar sprays. Not only are these more effective for pest control, but if applied in the root zone there are less run-off losses. Better, more specific, placing of the pesticide, reduction of drift to adjacent areas and the avoidance of excessive concentrations at the time of application are also preferable (Koeman, 1974), though all of these practices require the participation and understanding of the farmers. Spiller (1985) claimed that the variety of pesticides marketed in many countries further confuses the farmer into their misuse.

Table 7.5. Toxicity of six organosynthetic pesticides to common carp (*Cyprinus carpio* L.). (Source: Koesoemadinata, 1980.)

Group	Trade name [a]	Common name	Median lethal concentration (LC_{50}) [b] 24 hr	48 hr	96 hr
Organochlorides	Endrine 19·2% EC	endrin	0·0058 (0·0051–0·0066)	0·0049 (0·0044–0·0054)	0·0040 (0·0036–0·0044)
	Thiodan 35% EC	endosulphan	0·024 (0·021–0·027)	0·018 (0·014–0·022)	0·0092 (0·0087–0·0098)
Organophosphates	Nogos 50% EC	dichlorvos	3·80 (3·52–4·10)	2·70 (2·41–3·02)	2·30 (2·04–2·59)
	Folithion 50% EC	fenitrothion	6·00 (5·50–6·60)	5·40 (4·70–6·20)	3·40 (2·70–4·20)
Carbamates	Lannate 90% WP	methomyl	10·20 (8·80–11·60)	9·50 (8·26–10·92)	5·80 (4·90–6·80)
	Sevin 85% SP	carbaryl	31·50 (29·20–34·00)	14·00 (12·70–15·40)	8·20 (7·20–9·30)

(a) The active ingredient content indicated by % figures following trade names. EC: emulsificable concentrate; SP: soluble powder; WP: wettable powder.

(b) Values expressed in ppm formulated products, with 95% confidence limits.

Table 7.6. Toxicity and persistence of various pesticides used in rice fields to fish and laboratory rats. (Source: Singh *et al.*, 1980.)

Pesticide	LC_{50} to fish (ppb)	LC_{50} to white rats (ppm)	Persistence in the environment	Persistence in biological tissues
Thiodan	0·31–8·1 5–8 (24 hr)	100	Rapidly degraded, rapid but variable rate of disappearance	Degraded and excreted
8–BHC	22–53 (48 hr) 77–790 (96 hr)	125	Variable, degradable with rapid to slow disappearance	Persistent
Endrin	0·5–0·3 (48 hr) 0·6 (96 hr)	17·8	Persistent	Persistent, non-accumulated
Sevin	2000–25 000 (48 hr) 50 000 (96 hr)	–	Rapidly degraded	Non-persistent
DDT	5·0 (48 hr) 16·0 (96 hr)	113	Persistent	Persistent
Malathion	79–86 (48 hr)	1375	Rapidly degraded	Non-persistent

Special problems are encountered for pesticide use in rice cultured with fish. Since most rice pest control requires at least 3-4 applications in one growing season, rice culture can be severely curtailed if treatment is restricted. Draining of the water, and harbouring of the fish in central or peripheral trenches may avoid the worst effects, or complete removal of the fish during application may be necessary. Larger fingerlings are also known to be more resistant to pesticides and so their use is to be recommended. Field studies using carbofuran have shown zero mortality for fish released seven days after broadcasting or root zone application, compared to 100% mortality for paddies dosed when fish were present. Estores *et al.* (1980) also reports that carbofuran, as a water soluble compound, is not accumulated in fish tissues or persistent in crops or soil. As a result, carbofuran is the only pesticide recommended for rice/fish culture in the Philippines.

CHAPTER 8
HUMAN WASTES IN FISH CULTURE

8.1 Introduction

Water, domestic wastes and health are inextricably linked, especially in hot climates. Edwards (1980[b]) reported that approximately 25000 people per day die from preventable water-borne diseases, directly, or in conjunction with malnutrition. The proportion of the urban population within the tropics or under-developed countries which are served by sewage or water-borne systems has actually decreased from 27% in 1970 to 25% in 1976 whilst another 25% of the urban population has no access to any sanitary facility at all.

The introduction of adequate sanitation into a community may not be easy. The high financial cost of many systems may deter its introduction, as may social attitudes. The use of human wastes for fish production, and the income thus generated may, however, create the financial incentive for human waste to be collected and recycled (McGarry, 1977[a]). An assessment of aquaculture as a component of low cost sanitation has been produced by Edwards (1985[d]).

It has been estimated that the production of human excreta in Asia is about 13×10^6 t dry wt/yr, containing $9{\cdot}7 \times 10^6$ t nitrogen, $1{\cdot}4 \times 10^6$ t phosphorus and $1{\cdot}9 \times 10^6$ t potassium (McGarry, 1977[a]). Recycling such wastes would reduce pollution and insanitary disease if handled and treated properly. Furthermore, if converted into fish flesh, it could increase food production and help alleviate malnutrition. Schroeder and Hepher (1979) stated that although aesthetic considerations limited its use, human waste was equal to or superior to other animal manures, in increasing fish production. In addition, it has been reported that by holding fish in clean water ponds for several weeks at the end of the growing season, any objectionable odour and/or pathogens are obviated, and the fish become acceptable for marketing (Allen and Hepher, 1979). Stocking of fish in sewage-fed ponds also appears to aid the treatment process as these ponds have been observed to have higher pH and concentrations of dissolved oxygen than similar ponds without fish (Schroeder, 1975).

8.2 Human wastes collection—the alternatives

Human waste disposal by means of conventional treatment is often unsuitable for Third World countries, and incidentally, for integration with fish culture. Western-style systems are expensive; a capital cost of US$ 500, per household has been suggested (McGarry, 1977[b]). The large amounts of water required for these systems make them particularly unsuitable for many areas of the tropics; piped water is an expensive luxury available to only an estimated 13% of the population (Edwards, 1980[b]).

The large dilution of human wastes in conventional treatment systems (99·9% water) (McGarry, 1977[a]) may also reduce the efficiency of nutrient removal and benefit to fish culture. Moreover contamination by other domestic and even industrial wastes may also be difficult to avoid in sewered systems. McGarry (1977[b]) has reviewed the available methods of human waste collection making distinctions between their relevance for urban and rural application (Table 8.1). The World Bank has placed emphasis on developing sewerless methods, i.e. without water-flushed disposal and collection to tackle the sanitation problems of the Third World (Dale, 1979).

If aquaculture is to be integrated into the disposal of what are essentially 'dry wastes' (Rybezynski *et al.*, 1978) commonly known as 'nightsoil', the choice of collection method is important. The basic differences between 'dry' waste and sewage utilisation for fish culture are summarised in Fig. 8.1.

8.3 Human waste composition

The solid and liquid components of wastes produced by humans vary for many of the same reasons as those of other animals. The composition of solid faecal matter, however, remains essentially unaltered (Tacon, 1978). However, large differences are found in output of solid wastes for diets primarily containing meat and those containing vegetables, presumably this is due to a higher fibre content of the latter. Since heterotrophic production on fibre is an important food pathway for fish, such differences may have relevance in the estimation of loadings of human wastes in fish ponds. The faeces consist primarily of the residues of bile, and other intestinal secretions, mucus, leucocytes, desquamated epithelium and large numbers of bacteria.

Human urine is more variable in terms of its daily production and composition than solid wastes (Tacon, 1978). The approximate quantities and composition of the solid (faeces) and soluble (urine) fractions of human waste excreta are shown in Table 8.2.

Dilution of human wastes occurs when collected by water-borne, or sewerage, methods. According to Mara (1977) domestic sewage is composed of human excreta and sullage (waste water from the house, e.g. from sinks, baths, etc.). Municipal sewage may also contain flood water causing further dilution and by-products of trade and agricultural wastes. The quality and quantity of 'dry wastes' collected by non-sewerage means will also vary considerably. Losses through design and accidental seepage and soakway will reduce the quantity and tend to produce a more concentrated waste for fish culture, for example, septic tank sludge. With storage, qualitative change will occur as with animal wastes (section 3.5).

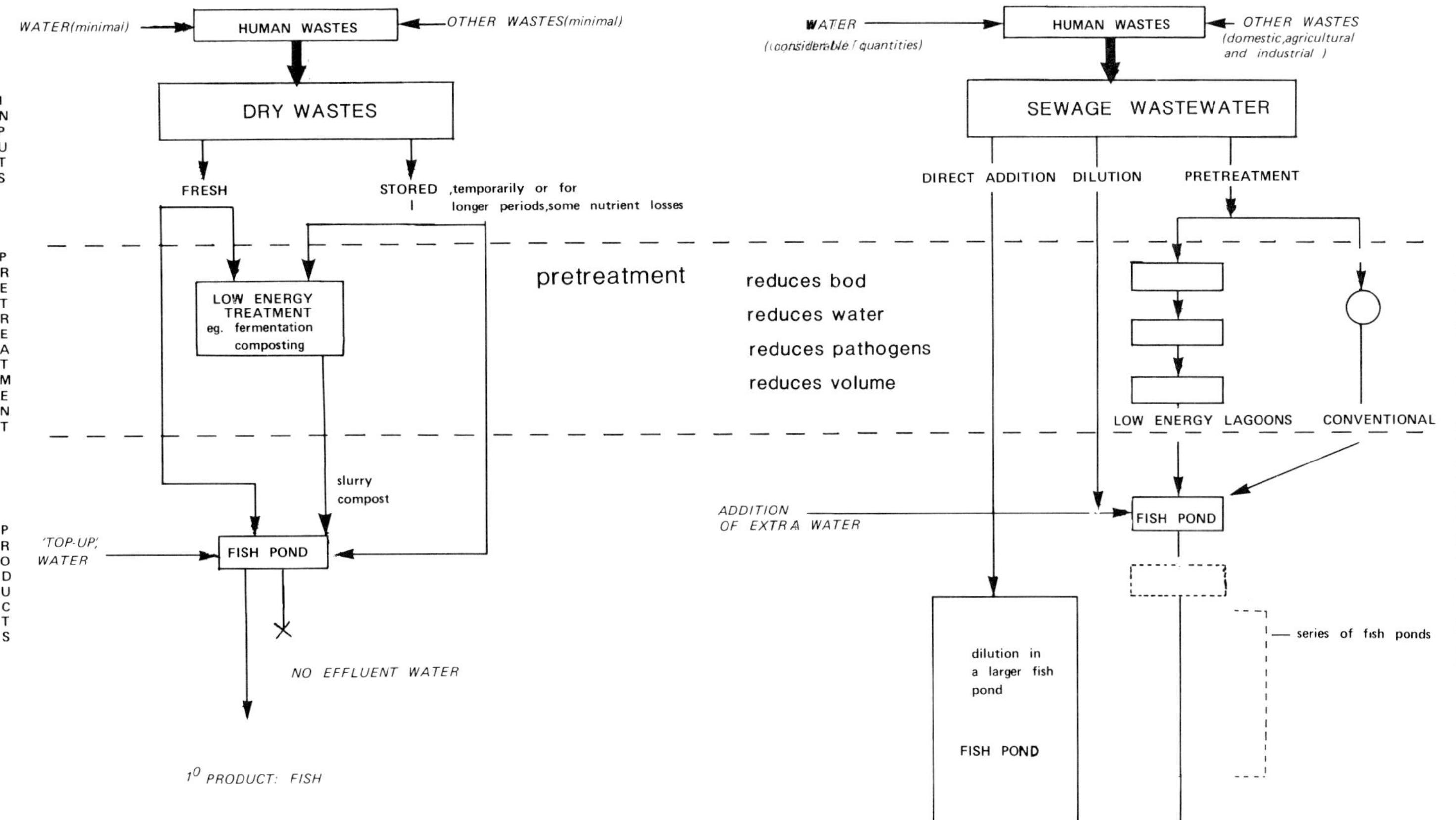

Fig. 8.1. Flow diagram comparing the use of sewage/waste waters with 'dry' human wastes in aquaculture.

Table 8.1. Summary of human waste collection systems and factors affecting integration with aquaculture. (After McGarry, 1977[a].)

Factors affecting aquaculture	Conventional sanitary sewerage (A)	Septic tank (B)	Vacuum truck and vault (C)	Aqua privy (D)	Compost toilet (E)	Bucket latrine (F)	Pitprivy/ bored hole (G)	Prai (H)	Overhung latrine (I)
ECONOMIC									
Water requirement	XXXXX	XX	XX	XX	–	–	–	–	–
Construction									
– Cost	XXXXX	XXX	XX	XX	XX	X	XX	XX	–
– Ease[1]	X	XX	XXX	XXX	XXXX	XXXXX	XXX	XXX	XXXXX
Collection									
– Cost[2]	X	XX	XXX	XX	XX	XXX	–	–	–
– Ease[3]	XXXXX	XXX	XX	XXX	XX	X	–	–	–
HEALTH									
Degree of sanitation	XXXXX	XXXX	XXXX	XXX	XXX	XX	XX	XX	X
Ability/ease of pretreatment	XX	XXXX	XXX	XXXX	XXXXX	XXX	X	X	–
CONTAMINATION									
Dilution by water	XXXXX	XX	XXX	XX	–	X	–	–	–
Other domestic or industrial wastes	XXXX	X	X	X	XX	X	XX	XX	–

ACCEPTABILITY – of use[4]	XXXXX	XXXX	XXX	XX	XX	X	X	X	X
– of waste collection[5]	XXXXX	XXXX	XXX	XXX	X				
INTEGRATION – Efficiency[6]	XXX	XXX	XXXXX	XXX	XXX	XXXX	X	X	XXXXX
– Potential[7]	XXX	XXXX	XXXXX	XXXX	XX	XX	–	–	XX

Notes:

A–F = urban systems; D–I = rural systems; A–D = human waste + water; A = human wastes + additional chemical/industrial contamination.

1. Ease of construction/implementation – depends largely on flexibility, i.e. if capable of incremental implementation amount of excavation required, suitability for average location
2. Cost in terms of labour and/or machinery/vehicles, etc.
3. Ease – often related to volume of wastes collected, degree of handling required.
4. General acceptability of system to be introduced – dependent on current social and communal practice.
5. Reflects whether the system is aesthetically acceptable.
6. Efficiency of nutrient collection and use by aquaculture. Systems allowing loss of liquid fraction show incomplete collection as do those employing large amounts of water.
7. As a result of the above factors.

XXXXX ——> X MOST ——> LEAST (– NONE)

Table 8.2. Characteristics of human excreta. (Source: Feacham *et al.*, 1977.)

Total wt/day	–	1·5 kg (2% body weight)
% Faeces	–	20
% Urine	–	80
% Water	–	80-90
Total solids	–	0·13 kg
N_2	–	0·008 kg/day
C:N ratio	–	6:1
BOD	–	0·08 kg/day

8.4 Dry wastes/fish culture

8.4.1 General considerations

The use of 'dry' human wastes as a traditional input into fish ponds occurs only in a few countries. The practice is only widespread in China where, as a valued input for agriculture generally, it is collected in a systematic way; latrines over-hanging ponds are also a common sight in parts of South East Asia.

8.4.2 Methodology

McGarry (1977[b]) has identified many of the problems associated with implementation of both rural and urban sanitation. The social and cultural basis of the community must be central to the design of collection and removal methods. The systems described (Table 8.1) for dry wastes show a variable ability for collection and storage. Commonly, the liquid or supernatant fraction is run to waste, with consequent loss of nutrients and risk of pollution. The vacuum truck and vault system has been suggested by several authors, based on the Japanese experience, to be an effective means of total human waste collection and removal. In this system excreta accumulates in concrete privy vaults and is collected every 3-4 weeks by small tankers with vacuum pumps (Pradt, 1971) (Fig. 8.2). The use of water after defaecation, whilst common in Asia, is not universal in Africa, for instance. The use of water-seal toilets is therefore problematic, and would restrict the use of vault and vacuum and aquaprivy methods for storage and collection.

Where waste collection is not on a daily basis, some form of storage is necessary. Large storage devices, such as septic tanks, only effectively retain solids, as the liquid fraction seeps away. This has consequences for the quality of the wastes. The processes for pre-treatment of 'dry' wastes before use in aquaculture, whilst fulfilling a similar function to those for 'wet' waste, are different in practice. The smaller proportion of water may encourage the use of composting techniques and biogas production for instance; the resultant compost and biogas effluent slurry are valuable pond inputs (see section 9.3.4).

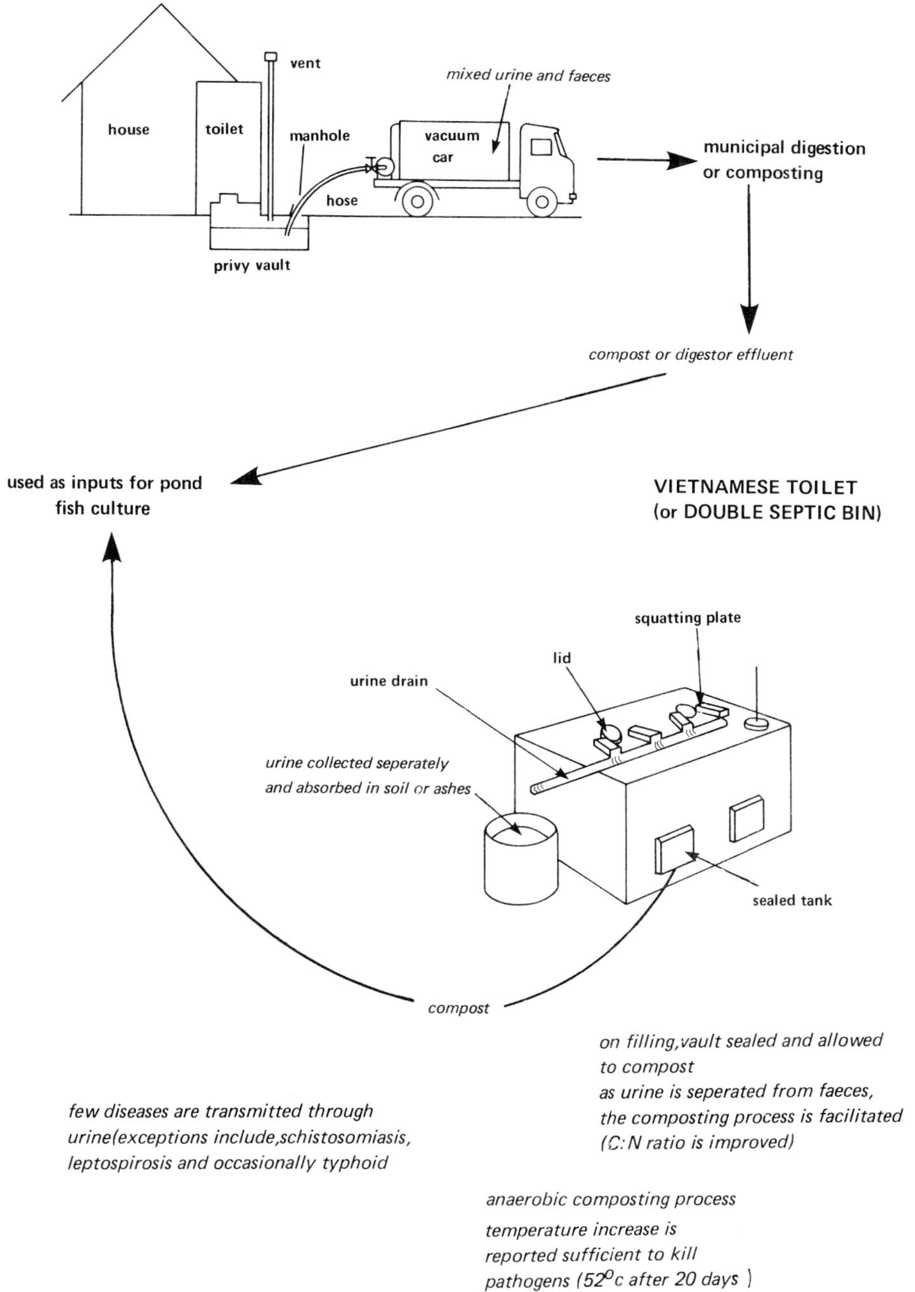

Fig. 8.2. Use of 'dry' human wastes in fish culture after urban and rural collection.

8.5 Sewage/fish culture

8.5.1 General considerations

Most of the integrated systems of sewage or wastewater treatment currently used were not designed to optimise fish production. The major priority was that of minimising construction, land and operating costs whilst achieving a given quality of sewage treatment (McGarry, 1977[b]). Certain authors have therefore recommended that if costs allow, sewage systems could be designed to maximise fish production. In general, this would necessitate using larger amounts of land and labour, which is not normally a limitation in less developed countries. Retention times of wastes within the pond would increase to improve nutrient uptake into the pond's foodwebs and hence increase fish production. Consequently the final quality might be expected to exceed that possible in conventional sewage oxidative ponds. Design criteria for a sewage-based aquaculture/agriculture system in Thailand are discussed later (section 12.3.4).

8.5.2 Methodology

The use of sewage wastes in aquaculture is similar to that of any other organic waste. Three approaches for supplying organic load into fish ponds have been defined (Allen and Hepher, 1979) (see Fig. 8.1).

(a) Pre-treatment

Sewage treatment reduced the BOD and pathogen loading before its use as a fish culture input. Treated sewage, whether by conventional means (trickling filter, activated sludge, etc.) or lagoon systems, has a lower nutrient content and BOD, and is thus more manageable as a fish pond input. As for other organic wastes, overloading the pond's capacity will result in imbalance, and fish mortality.

(b) Dilution

Wastewater is diluted with a varying volume of freshwater prior to its introduction into the fish pond. This again reduces the BOD and nutrient content.

(c) Direct application

The wastewater is given only a short primary treatment in settling tanks before direct addition to the fish pond. Since dilution of the sewage occurs within the fish pond, a large volume is required to absorb and digest the organic material.

Hepher and Pruginin (1981) reported that reduction of pathogens and toxicants to a level which would not endanger fish and/or fish consumers could only be reached by pre-treatment of wastewater. Consequently, they suggested that the dilution method should only be used with weak (usually rural) wastewater.

Edwards *et al.* (1980) have proposed a fish culture system, based on the use of septic tank sludge added to a single stage stabilisation fish pond.

8.6 Benefits of human waste/fish culture integration

The benefits derived from the utilisation of human wastes may be considered under two categories:

— those related to fish culture
— those related to human waste treatment.

8.6.1 Benefit to fish culture

According to their nutrient composition and the BOD they exert on receiving pond waters, human wastes should be comparable in stimulating fish production with any unadulterated animal waste. In general, however, fish yields in human waste-fed systems are inferior to those of animal/fish systems (see Table 8.3).

Table 8.3. Fish production and feed conversion factors for commercial Israeli fish ponds under various fertilisation schemes. (After Wohlfarth and Schroeder, 1979.)

	Feed	*Waste*
Yield (kg/ha/day)	50	32
Food conversion ratio (FCR)	2·5	2·7 (dry matter)
Cost/kg of input	3·0	0·3 (dry matter)
Cost/kg of fish	7·5	0·81
Total cost/ha/day	375	26

EXAMPLE 1 Computing the total cost of production of fish per day with conventional feeds and animal waste inputs.

	Feed	*Waste*
Yield (kg/ha/day)	50	32
Food conversion ratio (FCR)	2·5	2·7
Weight of input required	125	86·4 (dry matter)
Price of fish/kg	20	20
Total income/ha/day	1000	640

EXAMPLE 2 Computing the total income per hectare per day gained from producing fish with conventional feeds and animal wastes.

Assumptions: price of feed and fish are constant and price of manure dry matter is one tenth the price of the feed.

This is particularly the case with wastewater/sewage systems, normally because fish production is secondary to waste treatment. Also, many such systems have been situated in temperate, less productive regions. Possible growth limiting or toxic substances within sewage may also result in chronic low yields. Since pathogen levels limit the amount of sewage that can be used with safety, loading may also be lower than that required to optimise fish yields. Sewage-fed systems often used diluted effluent as the sole organic input. This contrasts with most productive animal/fish systems that use very high organic loadings and often receive high quality feed residues in addition, or even supplementary feeding.

Hepher and Schroeder (1977) reported the use of wastewater to improve food conversion in supplementary-fed fish ponds. A 40% reduction in supplemental feed, and 70% increase in yield was achieved over inorganic fertilisation (Table 8.3).

The use of 'dry' wastes, generally as nightsoil, has produced high yields. Edwards (1980[b]) quoted yields ranging between 9200-11 000 kg/ha/yr in fish ponds receiving human wastes, and Tang (1970) reported fish yields of approximately 7000-8000 kg/ha/yr in ponds having nightsoil as the main input.

Box 8.1. Theoretical loading rates of human sewage.

Typically, 45 g dry solids/day, 6·0 g N/day (see Table 8.2). A small family, say 2 adults/2 children (50% each) would supply ~135 g solids/day, (15·0 g N/day).

Compared with animal wastes, e.g. nitrogen, this is equivalent to ~4 pigs, or 2 cows

This would be sufficient for ~ 300–500 m^2 of pond. A 300 m^2 pond, with 5 t/ha/yr fish would produce 150 kg fish/yr which is equivalent of 37·5 kg/person/yr, which is similar to the highest world fish consumption levels. Now remember the practical problems!

8.6.2 Benefit to waste treatment

Pond systems designed specifically for waste treatment operate on similar lines to waste-fertilised fish ponds, through bacterial and algal growth (section 1). The rates of addition of materials should not exceed the carrying capacity of the pond, to avoid the accumulation of materials, the creation of toxic environments and inefficient pathogen removal.

Maturation ponds, which are generally used as wholly aerobic secondary and tertiary stages of lagoon systems (Mara, 1977) are the most suitable locations for fish production. The destruction of faecal coliforms is directly related to retention time and enteroviruses are destroyed best at depths of 1·0-1·5 m (Malherbe and Coetzee, 1965), conditions that are also optimal for fish culture. The presence of fish theoretically provides a higher nutrient uptake rate within the ponds and hence a greater overall efficiency of treatment.

The stocking of fish in such maturation ponds, with longer retention periods is therefore advantageous. The high photosynthetic rates prevalent in waste-fed ponds also allow for high dissolved oxygen and pH values during the daytime.

8.7 Constraints to use of human and domestic wastes

Some serious questions have been raised as to the suitability and safety of human waste integration with fish culture. Given that waste collection and use for aquaculture is economically viable, are these benefits negated by public health risks, the possibility of contamination and/or socio-economic problems? Rational answers to these questions are complicated by the way the constraints may arise (see Fig. 8.3), in different circumstances. When 'dry' wastes are used in fish culture, for instance, contamination with detergents is less of a problem than in sewage-fed systems.

8.7.1 Public health aspects

Numerous authors have reported the possibility of fish acting as active or passive vectors in the transmission of human diseases, and believe this to be a major constraint upon the integrated use of human wastes with fish culture. However, Bryan (1977) in a wide ranging preview of the medical literature found only two references to illness caused by fish contaminated by human wastes. Velasquez (1980) pointed out that the transmission of the majority of disease-causing organisms depended largely on environmental conditions coupled with the release of the host at varying intervals. Conditions occurring in waste-fed ponds might be expected to favour such organisms. However, it has been reported that unlike warm-blooded animals, fish themselves are not affected by infections of certain pathogenic species, such as *Salmonella* spp., *Shigella* spp. and other enterobacteria. These organisms may, however, be capable of either multiplying in the gut, mucus and tissues of fish, or utilising fish as passive carriers (Allen and Hepher, 1979). Fish are also intermediate hosts in the life cycle of some helminth parasites, such as *Fasciola,* the liver fluke. If the fish are consumed raw or undercooked, man can be infected. Buras and Niv Duek (1985) have examined the effect of both inoculation and exposure of various pathogenic organisms on the blue tilapia (*O. aureus*) and common carp. In particular threshold concentrations above which bacteria, bacteriophage and polio LSc virus could penetrate different parts of the fish were found to vary considerably. Laboratory trials indicated that although exposure to large numbers of bacteria and bacteriophage resulted in penetration of tissues and organs, adequate depuration in clean water for eight days was effective for their removal. However, large numbers of *Salmonella montivideo* could still be removed from the digestive tract even after duparation, highlighting the need for careful evisceration and cooking.

The environmental conditions created by factors such as high pH and high oxygen levels within waste-fed fish ponds result in the inhibition of a large number of pathogenic organisms, however. Following the addition of municipal cesspool slurry to earthen fish ponds in Bangkok, Edwards *et al.* (1984) reported a rapid attenuation of bacteria and bacteriophage levels over a 24 hour period. After this time, irrespective of the amount of wastes added, levels of these microbes in water and sediments were similar to those in ponds receiving no

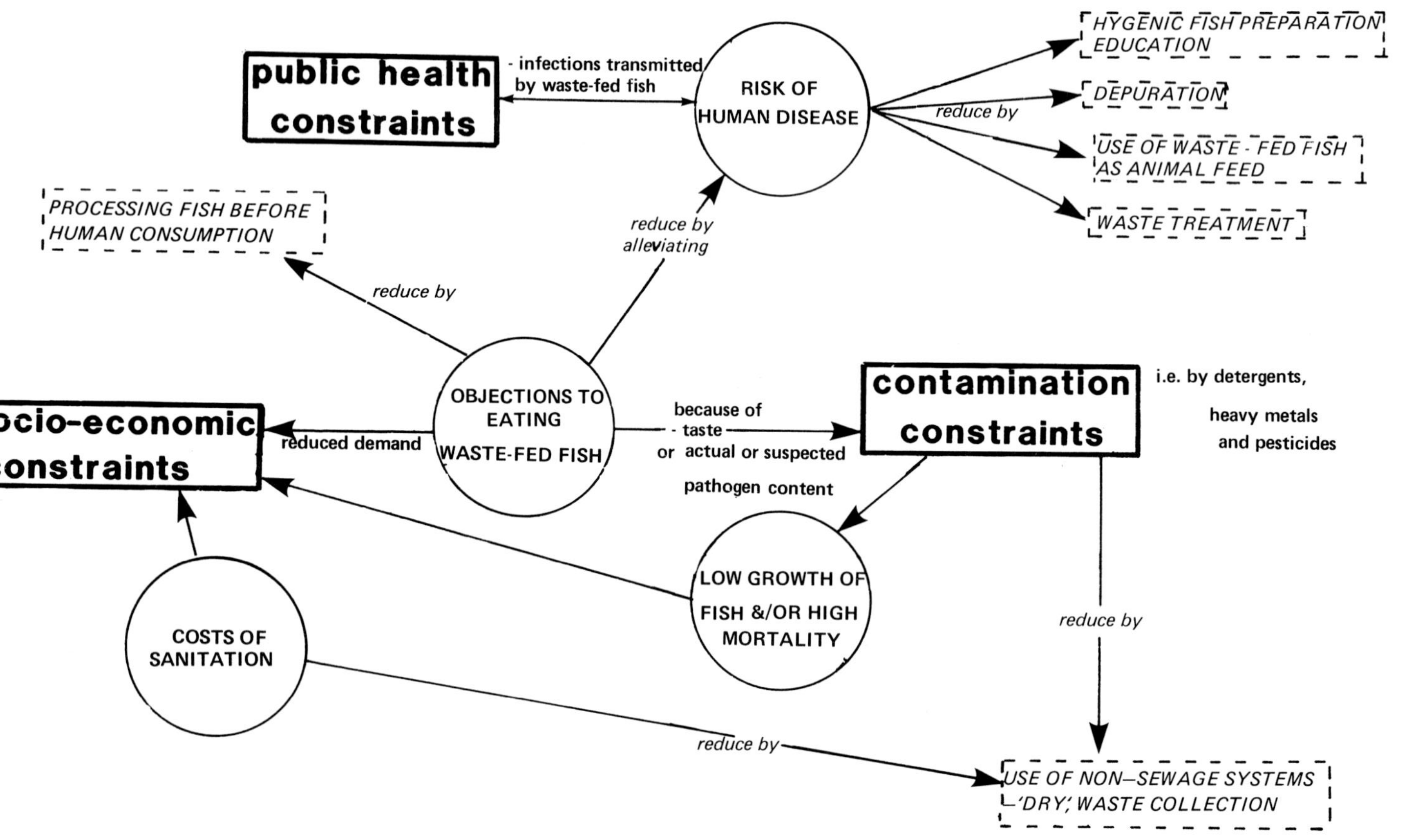

Fig. 8.3. Interrelationship of factors constraining the use of human waste in aquaculture.

wastes. It is therefore recommended that waste addition to ponds containing fish cultured for direct human consumption should be discontinued for a minimum of ten days prior to harvest. Alternatively the fish should be removed from the pond for depuration prior to sale. Fish should be immediately gutted after slaughter.

Certain measures may be taken to further alleviate the risk to public health, including pretreatment of wastes before use in ponds. McGarry (1976) reports that storage of nightsoil in sealed chambers in China has considerably reduced disease, and Edwards (1984) reported the absence of viable helminth eggs in cesspool and septic tank slurry, presumably because of the considerable period of storage. Similarly biogas production and composting both reduce pathogens by storage and heat.

The hygienic preparation of fish, removal of intestines and thorough cooking will undoubtedly reduce the health risk. In some areas, however, where it is customary to eat fish raw, partially preserved or ungutted, there may be a serious risk of increasing levels of infectious diseases. Improvement in traditional processing and preservation techniques wherever possible may be more successful than attempts to introduce radically new methods.

The use of waste-fed fish as a trash feed for other high valued species, or even conversion to animal feed would lengthen the disease chain to man and thus reduce the risks to human health. This approach is currently being tested in Thailand, where small, waste-fed Nile tilapia are being experimentally fed to catfish (*Clarias* sp.) and snakehead (*Channa* sp.).

Fish themselves appear to suffer little from diseases when raised on human wastes. Carp reared within wastewater ponds in Germany and Israel have been reported to suffer no more from disease outbreaks than fish reared within regular commercial ponds (Allen and Hepher, 1979). The same authors quoted preliminary studies carried out in Arcata, USA, which showed inhibitory effects on the virulence of *Vibrio* spp. in wastewater/seawater fish ponds.

8.7.2 Constraints associated with contamination of wastes

Hepher and Pruginin (1981) reported that detergents derived from domestic wastewaters, and concentrated in municipal sewerage systems or in water courses in rural areas, could be toxic to fish. Such chemicals, it is claimed, can affect the mucus coating and therefore render the fish susceptible to osmoregulatory problems, parasites and disease. These authors reported the lethal concentration of detergent for common carp at approximately 10 ppm of alkylbenzene sulphonate (ABS) and stated that higher concentrations of detergents, fluctuating between 10-18 ppm ABS for the wastewater of Haifa (Israel) have been recorded. The same authors reported that lower, sublethal concentrations of hard detergents could have chronic effects on fish growth.

Certain compound, e.g. phenols, although generally non-toxic, have been reported to produce off-flavours and undesirable odours in fish caught in polluted areas (Allen and Hepher, 1979). By-products from the production of single-celled protein (Tacon, 1978) can also produce off-flavours. Lovell (1979) found that compounds responsible for the so defined 'earthy-musty' flavour in fish were apparently geosmin or chemically related metabolites, which could be either synthesised by aquatic micro-organisms (actinomycetes and algae), or obtained from natural waters. In addition, wastewaters in general may contain potentially dangerous agricultural pesticides and herbicides (see section 7.5).

These problems associated with the use of certain wastewaters apart, it has been shown that fish grown in ponds fed with well treated domestic wastes were equal to, or even superior in, taste and/or odour than fish cultivated in other ponds (Allen and Hepher, 1979). These observations have been corroborated on wastewater-grown catfish (*Ictalurus* spp) in USA, common carp (*Cyprinus carpio*) in Israel and Germany, and cut-throat trout (*Salmo clarkii*) in USA (Allen and Hepher, 1979).

Fish usually grow fast in waste-fed ponds and thus may avoid the accumulation of heavy metals and pesticides. Hepher and Pruginin (1981) reported that fish have been shown to accumulate heavy metals and organic compounds in as little as 70 days although levels remained below accepted limits. Such contamination may prove impossible to eliminate from wastewaters and may effect the viability of integrated culture without affecting fish production directly. Unusual or unpleasant taste will only reinforce public attitudes against the use of human wastes as inputs.

8.7.3 Socio-economic constraints

Human wastes are a very available resource, but their use as inputs to fish culture may still be limited by the absence of an acceptable means of collection and removal. 'Dry' waste methods of collection, for both urban and rural areas have been identified as a flexible and cost effective solution.

Although it has been demonstrated that waste-grown products (including fish) can safely be used for human consumption, the majority of people are usually reluctant to consume such products. The degree of public acceptance of waste-grown products is determined by socio-cultural conditions as much as actual risk. Edwards (1980[b]) suggests that Chinese and Indian people have few objections to eating fish reared on human wastes, whereas Malay and Thai populations immediately reject such a suggestion. An investigation into attitudes to waste-fed fish in Pathum Thani, Central Thailand, substantiated these earlier observations (Edwards *et al.*, 1983[b]). A large proportion (70-80%) of those farmers questioned had previously heard of using animal and human wastes in fish culture, but had rejected the idea of eating them; the major reason given was a belief that the practice was dirty and disgusting. Marginally more people thought worse of eating human waste-fed fish than animal waste-fed fish.

It must also be recalled that the majority of fish do not actually feed directly on the waste; but on the natural food developed as a result of indirect fertilising. Allen and Hepher (1979) stated that the most critical factor in obtaining public acceptance of wastewater grown fish was their content of human pathogens. On the other hand, Edwards (1980[b]) reported that discussions with people willing to eat waste-grown fish indicated that they were not concerned about disease transfer if fish intestines were removed and thorough washing and cooking of fish flesh prior to consumption took place.

Box 8.2. Checklist for using human wastes in fish culture.

(1) Don't use fresh human waste in fish culture; pretreat by storage, fermentation or composting.

(2) Stop using waste around 10 days before harvest and consumption and/or depurate in clean water for 7–10 days.

(3) Always cook fish well before consumption; in areas with a tradition of eaten raw fish the method should not be promoted.

(4) Examine the use of waste-fed fish as a trash fish/fish meal source for other fish and livestock.

CHAPTER 9
PROCESSING AND PROCESSED WASTES

9.1 Introduction

Many of the inputs used in integrated fish ponds may be loosely described as 'processed' in some manner, that is their composition or characteristics have been changed in the course of their transfer to the pond. Animal manures are, in a sense, 'processed' plant material; domestic sewage may also be considered as processed as it consists of wastes that have been diluted.

This section, however, deals with products of the more conventional forms of processing or pre-treatment which are used as inputs into fish culture. Processed substances as used in fish culture can be categorised as:

— those derived from *crop processing* – the residues and by-products from the processing of 'raw' crops
— those derived from *waste processing* – the residues and by-products from the processing of liquid or solid wastes (see Fig. 9.1).

9.2 Crop processing wastes

9.2.1 Incentives to use processing wastes in fish culture

Agricultural crops are bulky and, in order to minimise transport and distribution costs, some form of processing close to their site of production is desirable. Thus, there is a natural incentive to reuse processing wastes, or rather by-products, within the agricultural system. These factors are especially pertinent when the proportion of by-product as a percentage of the original crop is considered; for example every tonne of dehulled coffee produced results in 4·5 tonnes of residues. Typically, processing will yield a variety of products, e.g. the milling of rice produces rice hulls, bran, broken rice and polished rice. These fractions have different economic and feed values, and may be more or less sought after within the agricultural sector for livestock feed and other purposes. Often, processing also necessitates the use of large volumes of water (e.g. up to 6000 litres per tonne of processed coffee berries). The resultant effluent, if untreated, imposes a high BOD on receiving waters and can create a serious pollution problem. Rubber and oil palm industries for instance, are responsible for widespread pollution in Malaysia (Ma, 1975).

Where physical arrangements allow, fish culture can use many agricultural process wastes. Large-scale plantations and farms have been increasingly diversifying into processing and use of produce on the same site, especially

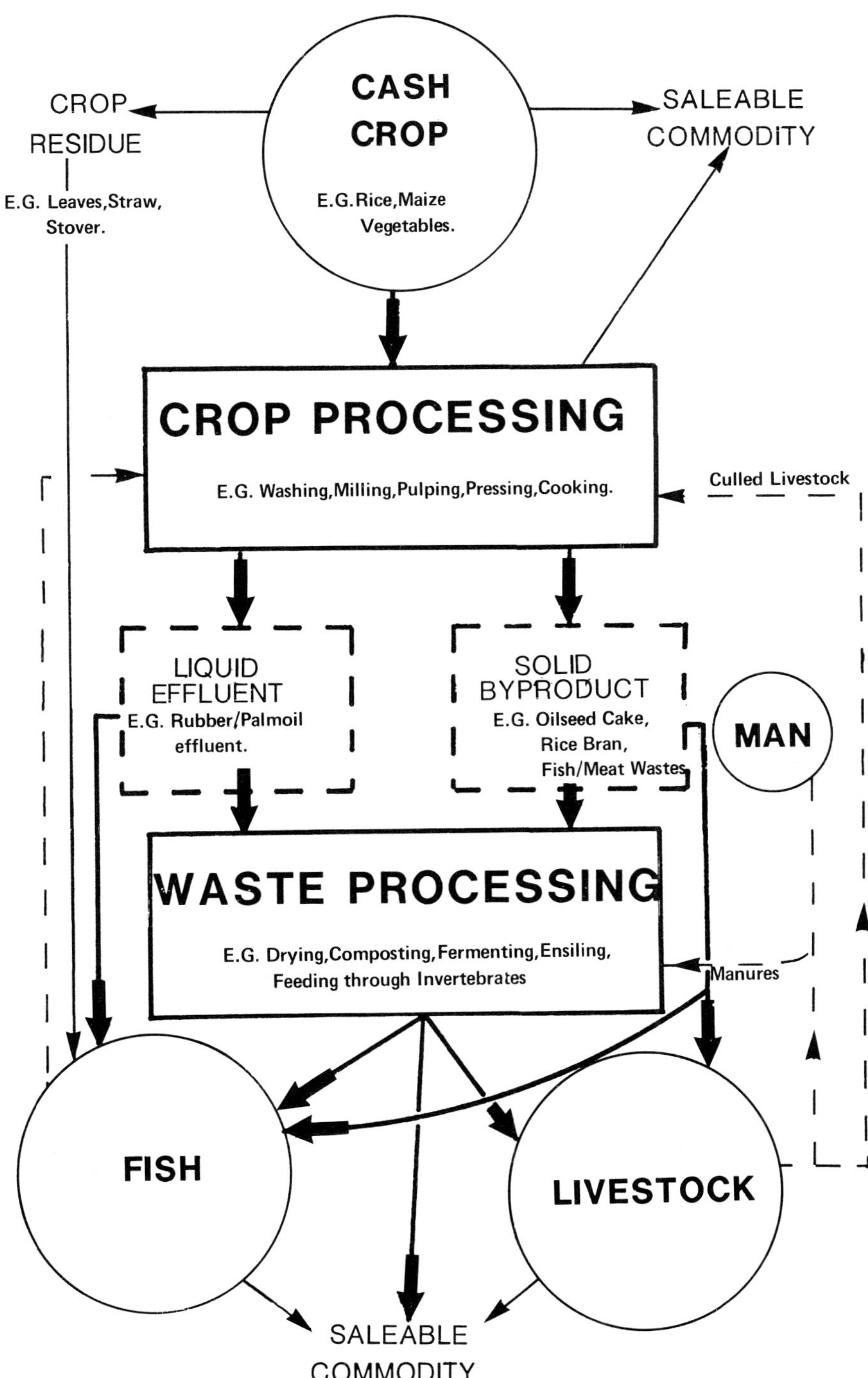

Fig. 9.1. Crop and waste processing before use in fish culture.

where the profits from cash crops are low or variable. Processing on a smaller scale can also be attractive, especially for locally grown and used products such as rice and sago. Processing wastes from sugar cane grown intensively on fish pond dykes have been used as traditional pond inputs at rates of up to 60 t/ha/yr in China (Ruddle, 1985[b]). The widespread introduction of small diesel-run cassava choppers, rice mills, etc. into rural areas creates both employment and by-products for local recycling. The value and cost of processing wastes is not only dependent on their nutrient or feed value. When concentrated wastes are produced in isolation, that is, without a diversified local economy to absorb and use them, they remain wastes. Conversely the same product in an area of livestock production might be considered of great value, and its use in fish culture may be uncompetitive.

9.2.2 Processing wastes used in complete feeds

Many processing by-products have a high feed value and their use simply as supplementary feeds may be wasteful. The development of complete feeds from combinations of such by-products and other more conventional materials has become a priority in some sectors of fish culture. The rationale of this approach is not only to allow more intensified production of fish but also to minimise feed costs and wastage (FAO, 1983). The preparation of balanced, complete diets, from a variety of unconventional sources has proven potential on a large centralised scale, but is not likely to be practical on a smaller, local scale. The centralised production of complete feeds, even if based on low cost by-products tends to lose many of the advantages that may be achieved by using simple integration locally. Also, centralised production of fish feeds is likely only where the technical facilities and expertise are already established in the form of an animal feed industry. The use of complete diets also assumes a high market value for fish; any relevance for subsistence systems is therefore unlikely.

The use of processing by-products therefore is considered only in terms of their value as supplemental feeds, rather than in their inclusion in complete diets. An excellent account of their use in complete diets for fish and a country-wide survey of fish feed processing industries is available (FAO, 1983). The use of animal processing wastes in fish culture is not included, since they are rarely used as supplementary feeds. (Table of processing by-products, Appendix 10.)

9.2.3 Variation in feed quality of processing wastes

A wide variability in feed quality (as estimated by crude protein, lipid and carbohydrate) exists amongst by-products of the same type (Appendix 8). This is to be expected since the raw materials also exhibit considerable variations in quality and composition (section 5.4.1.). Climate, variety of plant and method of cultivation all contribute to regional and local variability.

Even more important may be the method and extent of processing. This is particularly the case with oil-cakes which are the residues of a variety of oil-extraction methods. Hydraulic pressing is the principal method of oil extraction in most of Asia (FAO, 1983) and gives a higher quality cake than the screw-expeller method that uses steam heating; its inefficient extraction results in a cake of higher lipid content being available as a feed. This production method also aids in the removal of toxins present in the oil-cake (Appendix 6).

The quality of protein, as indicated by the presence and quality of essential amino-acids, is important when by-products are used as part of complete feeds. When by-products are used only to supplement natural feeds present in fish ponds, such a sophisticated balance of feed inputs is unnecessary.

9.2.4 Liquid processing wastes

The large amounts of liquid processing wastes generated by agro-industry every year is an area of great potential for integration with fish culture. Provisional experimentation with rubber wastes (see below) have indicated yields in the order of those for sewage-fed fish culture, without the hygienic constraints already discussed (section 8.7.1). Undoubtedly these results warrant further research for other processing industries.

The quality of effluent from natural rubber processing varies depending upon the type of operation (John, 1975). Effluent consisting of uncoagulated latex, serum and wastewater (Yap *et al.*, 1983), is produced in small plants in rubber growing areas or in large and centrally located factories. Yap *et al.* have attempted to grow fish in the second and third ponds of an effluent treatment lagoon system. A monoculture (stocked at 5000/ha) of silver striped catfish (*Pangasius sutchi*) showed excellent growth with a production of nearly 7000 kg/ha/yr on an extrapolated basis. A polyculture including species sensitive to low levels of dissolved oxygen suffered high levels of mortality however.

Since the effluent quality and quantity associated with such industries is likely to be variable, periodic overloading of fish ponds with wastes is a problem. Treatment of such liquid wastes prior to their use as pond inputs may improve their management for fish culture (section 9.3.4).

9.3 Waste processing

Use of processed wastes in fish culture

The difficulties and constraints associated with the use of fresh wastes have been discussed. Further processing may improve the waste's value, or its acceptability for use in fish culture. The general advantages of using processed wastes in fish culture can be summarised as:

— reduction in volume and weight of the waste, resulting in a more concentrated feed

— production of a more homogenous substance from a combination of raw waste materials; generally this would improve their management as pond inputs

— reduction of pathogens and improvement in stability of a waste product; this would increase the acceptability of handling by the fish farmer and to the consumer.

The processing of wastes may also however, affect their use as pond inputs by:

— increasing their value so that it is no longer economically viable to use them in fish culture
— reducing the feeds value for fish compared to fresh waste (depending on processing method)
— allowing a more diversified and versatile use of waste within an integrated farm.

A summary of processed waste products and their use in fish ponds is given (Table 9.1).

Sections 9.3.1–9.3.5 deal with methods of processing wastes before their use as pond inputs. These methods include simple drying and various forms of organic decomposition (Fig. 9.2) and fermentation such as composting and ensiling.

9.3.1 Simple drying

Animal manures

The use of dried manures often marks the first stage in intensifying production from subsistence fish ponds. Faeces, often of free-roaming domestic animals, that have been dried by exposure to sun and wind, are collected regularly and simply added to ponds. Tacon (1983) records the use of sheep manures in this manner as inputs for subsistence carp ponds in Bolivia, and buffalo manure is used in the same way in Thailand.

Dried manure is often a traditional crop fertiliser or fuel source and as such is generally found more acceptable to handle and use than fresh waste. However, the nutrient content of dried waste, and thus its value as a fish pond input, is inferior after sun drying. Degani *et al.* (1982) found that sun dried, fermented cow slurry lost 60% of the total nitrogen, 75% of the phosphorus and 3·5% of the carbon. Nitrogen is usually considered more volatile and easily lost from the wastes than phosphorus (Taiganides, 1977) (section 3.4). In addition, the potential advantages of rapid dispersion of slurry and fresh manure into the water column will be lost by drying. Drying of slurries in particular, which have a higher water content than solid faecal waste, can also be problematic.

Plant wastes

Drying of plants and plant products at the site of harvest may, by removing some of their considerable moisture content, reduce costs of transportation and distribution. This is an important factor if large quantities are to be used. Simple wilting in the sun can reduce moisture content from 80% to 40% for instance (Leach, 1976), without any corresponding decrease in feed quality for ruminants. A potential reduction in palatability to fish may be a problem if the plant material is used as a direct feed for fish rather than a green manure. Thorough drying, to around 20% moisture, will prevent decomposition, and

Table 9.1. Summary of waste processing methods and product's usefulness as fish pond inputs.

Method	Degree of pretreatment handling	Acceptability of use	Methodology of processing: Cost	Methodology of processing: Skills	Nutrient feed value of product	Environmental effect	Competition for use of product	Pathogen attenuation
Drying	X	XXXX	X	–	X	Drying may be difficult during the wet season or in the humid tropics	Locally for fuel and fertiliser for crops	Good
Aerobic composting	XXXX	XXXX	X	X	Little fertiliser value. Appears potentially useful feed value for tilapia	Processing is more rapid at high temperature	Considerable as a soil conditioner and fertiliser	Very good if prepared properly
Anaerobic processing (a) compost	XXXX	XXX	X	X	Feed value not known	Processing is more rapid at high temperature	Considerable as a soil conditioner and fertiliser	Not good over short-term scale. Improves with storage
(b) digestor slurry/sludge	XX XXXX	XXXX	XXXX	XXXX	Improved feed value compared to unprocessed waste, greater nutrient availability	Processing is more rapid at high temperature – but nutrient content declines	Variable – but often high	Good – improved with storage or drying
Fish silage	XXX	XXXX	XXX	XXX	XXXXX	–	XXXX	XXXX
Invertebrates	Variable High	XXXX	XXX	XXX	XXXX	Moisture and temperature affect growth rate	Variable – likely to increase	Not known – probably good
Crop processing products	X	XXXX	Variable	Often high	XXXXX (Variable – see text)	May affect storage	Often very high	–

XXXXX HIGH X LOW

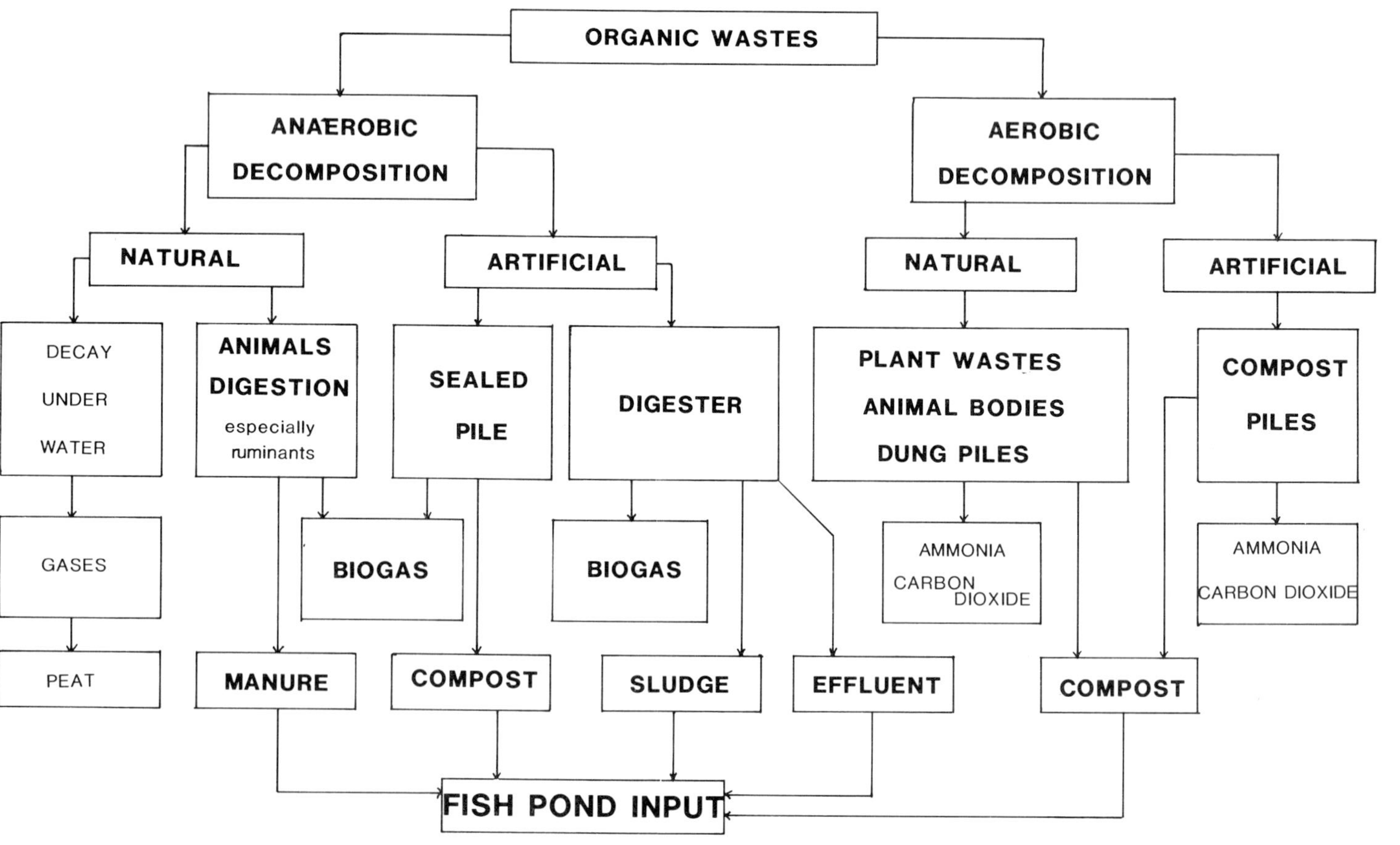

Fig. 9.2. Processing of organic wastes by decomposition.

facilitate storage of plant materials. In seasonal climates, or when the dried plant material is to be included as part of a complete pelleted diet, this may be advantageous.

Box 9.1. How much dried cow manure to equal wet weight?

Moisture content of wet cow manure = 79% i.e. 21% solids

Typical dried manure moisture content = 20% i.e. 80% solids

Thus 1 t dried manure = 800 kg solids

800 kg solids requires 800 × 100/21 – ~ 4·0 t wet manure

Note, however, that if pond loading rate is based on quality;

nitrogen loss is 60% on drying, thus for an equivalent nitrogen input 1 t dried manure has 40% of the N_2 equivalent weight in fresh manure. Thus 1 t dried manure ≡ 4·0 × 40/100 = 1·6 t fresh manure.

9.3.2 Aerobic composting

Edwards and Kaewpaitoon (1982) have identified compost as a low cost feed for subsistence rural fish production. Using simple methods it is claimed that significant yields of the Nile tilapia can be produced from readily available material such as water hyacinth, nightsoil and rice straw after composting. Normally ignored as a fish pond input because of a low nutrient content and hence poor fertilisation effect, recent evidence suggests that it may have value as a direct feed.

Methodology

The basis of aerobic composting is the decomposition of organic matter by bacteria. The number of bacteria increases using the nutrients in the waste as feed; as a result the amount of carbon in the feed material is reduced, mainly through respiration losses as carbon dioxide. If the amount of carbon originally contained in the materials to be composted is high, compared with nitrogen (a high C:N value), decomposition will be slow—since the bacteria require adequate nitrogen to grow rapidly. A C:N ratio of around 25:1, such as for fresh water hyacinth, is considered ideal for rapid composting. However, even materials with higher C:N ratios of up to 100/110:1 (e.g. rice straw) can be used for compost. The excess carbon is gradually reduced whilst nitrogen is recycled and conserved by succeeding generations of bacteria within the compost (Gotaas, 1956).

A mixture of animal manures (low C:N ratio) and low protein plant residues (high C:N ratio) will usually compost well, providing moisture content is adequate. Under tropical conditions, compost can be made in a matter of days or weeks if the C:N ratio is optimal. Heat is produced by the bacterial respiration during aerobic composting and temperatures may reach 45–65°C, a factor which makes this method a useful form of processing for unsanitary wastes. Conditions within a compost pile should remain aerobic for rapid and complete decomposition. Various means can be used to ensure this occurs depending on the scale of the process. Traditionally, compost is turned during the decay period, although this may be time consuming and can also cause heat to be lost. On a small scale ventilation has been ensured by insertion of bamboo poles in the pile (Fig. 9.3), whereas mechanical aeration has been used on a larger scale.

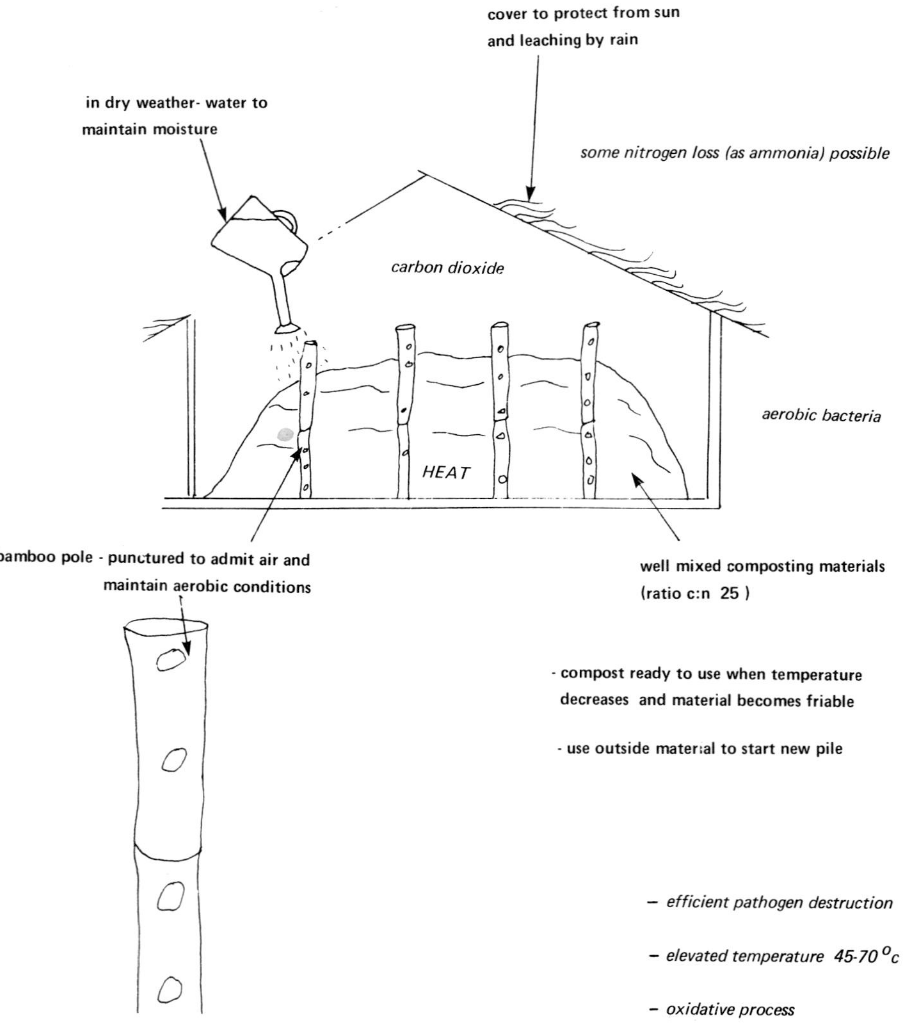

Fig. 9.3. Simple aerobic composting.

Use to aquaculture

(a) Feed quality

The aerobic composting process leads to some loss of nitrogen, in the form of ammonia, especially if the quantity of nitrogen is initially high. Phosphate and potash are more effectively conserved; carbon is lost in large amounts as carbon dioxide. The Chinese, who use compost especially for fry production, consider this an advantage, since the product has a decreased oxygen demand compared to the raw materials (Edwards, 1985[c]). The nutrient value of compost varies widely, depending mainly on the type of the materials used (see Table 9.2).

Table 9.2. Variation in compost quality. (After Gotaas, 1956.)

Substance	% by weight (DM)
Organic matter	25–50
Carbon	8–50
Nitrogen (as N)	0·4–3·5
Phosphorus (as P_2O_5)	0·3–3·5
Potassium (as K_2O)	0·5–1·8
Ash	20–65
Calcium (as CaO)	1·7–7·0

(b) Yields

In a series of experiments, Edwards *et al.* (1983[a], 1984) have studied the use of aerobically composted water hyacinth as a feed for the Nile tilapia (*O. niloticus*). A mean extrapolated yield of 3·6 t/ha/yr was obtained with relatively low feeding rates. Such yields may be explained by the possible ability of *O. niloticus* to hydrolyse the compost detritus as an energy source, rather than relying on the feed value of the highly proteinaceous microbes alone (Bowen, 1982, Edwards, 1982[a]). The use of composted water hyacinth in pelleted feed has also given promising results. Good growth and utilisation efficiency was achieved when up to 75% of the dietary material was replaced by compost (Edwards *et al.*, 1985). Yields of between 5-6 t/ha/yr (extrapolated) have been obtained in small ponds by adding fresh, chopped water hyacinth, allowing unchopped water hyacinth to decompose in the water, or by adding ready composted material. The same dry matter (200 kg/ha/day) and total Kjeldahl nitrogen (3 kg/ha/day) loading rates were used for each treatment. However, lower fish production was found in recent trials using compost made only from

Box 9.2. Making a good C:N mix.

If a variety of wastes are available for decomposition, mixing in the right proportion will ensure the best results. Animal wastes have low C:N ratios compared to vegetable waste. Thus for:

vegetable waste, e.g. rice straw C:N = 100:1

animal waste, e.g. chicken manure C:N = 9:1

The simplest way, with several waste components is to check the C:N ratio of what's available, then calculate as follows:

thus, with a mixture of 400 kg of cattle waste (@ C:N 15:1), plus 500 kg chicken manure (@ C:N 9:1). The final C:N will be

$(15 \times 4/9) + (9 \times 5/9):1 = 6{\cdot}67 + 5 \rightarrow 11{\cdot}67:1$

Remember to allow for the moisture content!

Alternatively where two components are used, a simple graph can be used as shown below.

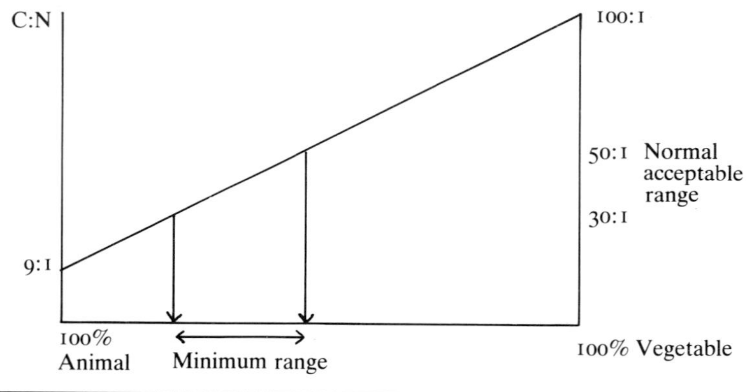

rice straw (Edwards *et al.*, 1984) suggesting that for the detritus consuming tilapias the C:N ratio of compost materials and the final compost may be critical to its use.

Strategies for compost production and use

(a) Human 'dry' waste disposal

Edwards *et al.* (1983[a]) have demonstrated the feasibility of disposing of nightsoil by composting with water hyacinth (C:N ratio 20) before feeding to fish. Gotaas (1956) suggests that a C:N ratio of above 30 is desirable when composting nightsoil in order to ensure adequate heating for pathogen destruction with a minimum nitrogen loss. As a guide to preparation he suggests that the ratio of nightsoil and other materials (binders) is approximately correct when a person is able to put his foot on the material without it sinking in. Pathogen

Box 9.3. Making and using water hyacinth compost for feeding the Nile tilapia at AIT.

(1) To make one pile of aerobic compost
Take 400 kg of sun dried ($\sim$1000 kg wet wt) water hyacinth and 400 kg of fresh water hyacinth. Mix well and place onto a thin layer of rice straw ($3{\cdot}5 \times 3{\cdot}5$ m). Pile to a height of around 1·2 m inserting perforated bamboo poles to ensure aerobic conditions.
Turn the pile after 14 days, and start to use after 45 days.

(2) Loading rates
To produce $\sim$55 kg of Nile tilapia in a 200 m^2 pond in 6 months feed compost at 4 kg/pond/day ($\sim$2 kg DM/day).

(3) How many piles required?
Dry matter loss during composting is $\sim$20%.
Therefore use $2 + (2 \times 0{\cdot}2) = 2{\cdot}4$ kg DM in preparation of compost.
Fresh wt equivalent of 2·4 kg DM is approximately 30 kg.
Since one pile contains 400 + 1000 kg wet wt, this quantity is enough for $\sim$46 days,

or 4 piles of compost are required for a 6 month culture period.

destruction by heat is an important advantage of aerobic composting. Most pathogens and parasites are rapidly killed by the temperature rising to 65-70°C inside the compost pile. Gotaas (1956) suggests that the outer, cooler layers are not so effective in reducing pathogens and recommends that compost heaps containing human waste should be subjected to occasional turning to ensure adequate and thorough heating. Alternatively, the outer layers can be used as a 'starter' for new piles.

(b) Human sewage sludge disposal

Parr *et al.* (1978) have reported the rationale and methodology for composting municipal sewage sludge before its use as a land application. Similar methods might be feasible to provide inputs for integrated tilapia culture. Compost can be produced in as little as three weeks using mechanically aerated piles. A potential constraint to its use as a fish pond input, in common with other sewage derived residues, is contamination with agro-industrial wastes (metals, detergents, etc.).

(c) Livestock production

Intensive livestock production where animals are raised on litter could produce composts for fish production. Intensive broiler production, for instance, is typically managed so as to require minimum amounts of litter and labour. The materials used for litter, such as rice husks and wood chippings, are less suitable for composts compared to straws and stovers because of their high silicon and lignin content which renders them only slowly decomposable. The resultant dry poultry litter is often not a good input to ponds. Use of more readily degradable litters and aeration may produce a better quality compost and also provide a more healthy environment for the growing birds. Overnight ruminant confinement might also be adopted for compost production on a small scale, if larger quantities of bedding material and more frequent removal is practised. Conditions of shade, and moisture from the animals' excrement, encourage a rapid change to compost.

9.3.3 Anaerobic composting

The main differences between aerobic and anaerobic composting are detailed in Fig. 9.2.

When used as a fish pond input, compost has been considered mainly as a fertiliser. Anaerobically composted material retains nitrogen more effectively than material composted by aerobic methods. Conditions within anaerobic compost piles, however, will often favour denitrification or the production of ammonia from the more stable nitrogenous compounds, nitrites and nitrates. Thus, the effective sealing of the pile, either with mud or water, is required to conserve nitrogen. Pathogen attenuation is much slower in anaerobic composts since there is no substantial release of heat. The pathogenic organisms have been found to disappear, over a period of 6-12 months, owing to the unfavourable environment and biological antagonism (Gotaas, 1956). This is a significant factor if nightsoil is to be composted.

Methodology

Anaerobic composting is reported to be widely used in China as a pre-treatment for nightsoil and other wastes before use in fish ponds. According to Delmendo (1980), during the composting of animal manures the raw material is collected and placed into circular composting pits (Fig. 9.4), which are filled by layering river silt (7·5 t), rice straw (0·15 t) mixture, pig or cow manure (1·0 t) and aquatic plants or green manure crops (0·75 t) in 15 cm layers. The top of the compost is covered with mud and a water column 3·4 cm deep, thus providing anaerobic conditions, and left for one month when the contents of the pit are turned over. Superphosphate (0·02 t) is then added, and thoroughly mixed in with water, so that moist conditions are ensured. The compost is turned again twice before using several months later.

The chemical composition of the compost is reported to be approximately N, 0·30%; P, 0·20%; K, 0·25%; organic matter 7·8-10·3% with a carbon: nitrogen (C:N) ratio of 15-20:1 (% wet weight).

In China, composted animal manure is applied to fish ponds three times during the fish culture period (six to eight months), the first application being greater than the last two and applied before fish are stocked in the pond (Delmendo, 1980). The amount added varied between five to over ten tonnes per hectare, depending on the type of soil. However, Hepher and Pruginin (1981) reported the construction of composting tanks so that the liquid product of fermentation flowed directly into the water supply canals and to the fish ponds.

Delmendo also reported the lower fertiliser potential of composted cow manure compared to the fresh waste due to denaturation of the protein content

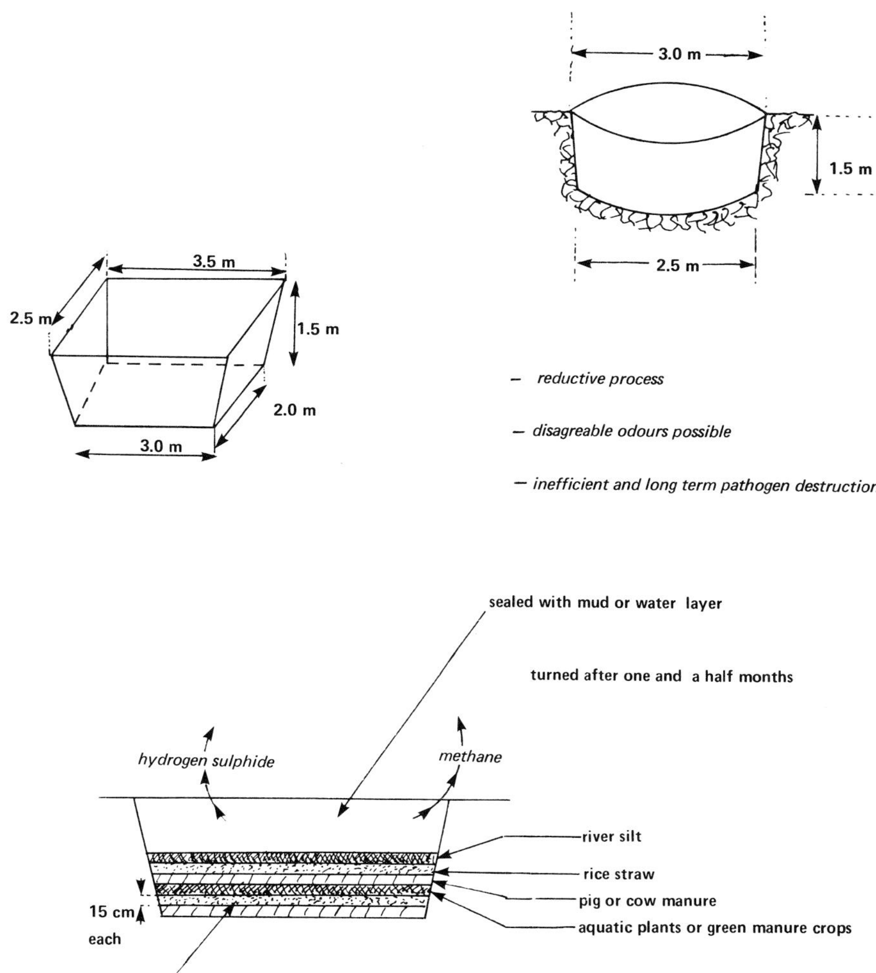

Fig. 9.4. Anaerobic composting in China.

and its subsequent binding into larger clumps which did not disperse well in the pond water, even after previous dilution.

9.3.4 Slurry and sludge by-products of methane production

Principles

Anaerobic digestion of organic waste produces methane gas which can be collected and used as a fuel and a slurry and sludge of value to fish culture. A sealed unit is typically used to digest the waste and collect the gas, and depending on design (see Fig. 9.5) may be constructed from a variety of materials. Anaerobic digestion to produce 'biogas' has been identified as a method for producing energy from waste and also a liquid effluent of value to fish production. Digestor slurry is already used as a fish pond input in South East Asia, especially in the Philippines (Crocker, 1983). In contrast to composting, using solid materials, feed materials for digestion are usually added in a liquid or slurry form. A sludge is produced within the digestor as well as the gas and slurry which is normally removed only irregularly. The slurry effluent is removed on a continual basis at a rate depending on digestor design and frequency of waste addition.

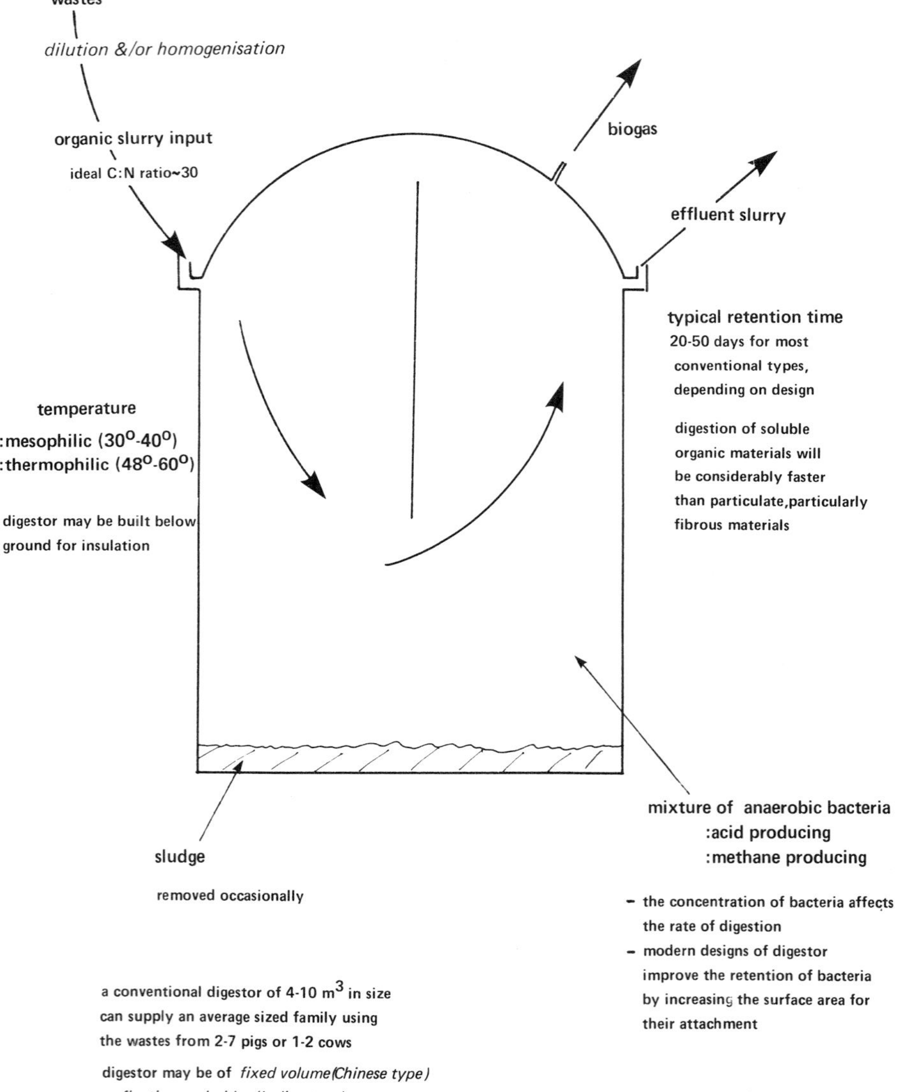

Fig. 9.5. Generalised design and operation of methane generator.

Operation of the digestor
The action of bacteria within a digestor is basically the same as for anaerobic composting. However, since the methane producing (methanogenic) bacteria are sensitive to oxygen, anaerobic conditions must be strictly maintained.

Digestor design
Digestors are basically of two types: batch loaded, and those loaded continuously or semi-continuously. Batch loading digestors, although simpler, tend to require a longer retention time to digest a given quantity of input material and the capital cost is likely to be high. Continuous digestors are more generally used.

Waste quality, loading and dilution rate
Although most forms of fresh, non-woody biomass may be digested to produce biogas and a valuable slurry for fish culture, the process design, yield and composition of the gas are affected by the type of material. The physical form of feed, stock handling method and the proportion present as an indigestible fraction must all be considered. The suitability of input feed can roughly be estimated by the C:N ratio, which as with aerobic composting has been determined as being best in the vicinity of 30 to 1 (McGarry, 1977[b]). Digestion of large amounts of fibrous vegetable material can be difficult practically, especially in continuous flow-type digestors, because of blockages and crust formation. Homogenous and liquid wastes are more effectively used, and pretreatment dilution and maceration can be carried out to achieve this. Typically feedstock for continuous flow digestors contains 3-10% solids.

Operating temperature and pathogen destruction
Two ranges of temperature are suitable for the digestion process: *thermophilic* (48-60°C; optimum 53°C) and *mesophilic* (30-40°C, optimum 35°C).

Despite the longer retention times that are required, use of mesophilic fermentation is recommended; thermophilic bacteria are most sensitive to environmental changes and the sludge by-product may have an unpleasant odour and be of poor fertiliser quality.

Pathogen destruction however is likely to be most effective with high temperatures. McGarry (1976) suggests that short retention times can be a source of trouble in this connection, and suggests that a period of more than 14 days at above 35°C is required to remove most pathogens. In practice, most digestion occurs under mesophilic conditions, unless additional heat is used (Table 9.3).

Use to aquaculture

(a) Fresh slurry
The direct benefits to aquaculture are realised through the feed and/or fertiliser value of slurry and digestor sludge. Degani *et al.* (1982) have demonstrated in

Table 9.3. Kill rates of pathogens during anaerobic fermentation. (After Barnett *et al.*, 1978.)

Organism – disease	Temperature °C	Retention time (days)	Kill rate (%)
Salmonella typhosa	22–37	6–20	99
Myobacterium tuberculosis	30	n.a.	100
Ascaris lumbricoides	29	15	90
Polio virus–l	35	2	98·5

tank systems the potential value of a slurry produced by thermophilic digestion of cow manure. The production of the blue tilapia (*Oreochromis aureus*) fed a diet of 50% digested slurry and 50% low protein feed was similar to that of fish fed with high protein pellets. A measured increase in the availability of algae in the pond water suggested that fish growth could be maintained by replacing the protein in feed by direct natural food production.

The growth of natural feeds stimulated by the addition of digestor slurry is most beneficial to fish such as the Nile tilapia which feed directly on phytoplankton. Schroeder (quoted by Degani *et al.*) found that digested cow manure (mesophilic), fed at 2·5% body weight/day, could replace pellets or liquid cow manure without affecting yields of tilapia in pond polycultures. Replacement of pellet feeds by slurry has been less successful with other fish, such as common carp, which do not feed directly on phytoplankton. Further, when slurry and pellets are used together at different ratios in ponds, Barasch and Schroeder (1984) found that fish yields became more variable as the proportion of pellets decreased. Thus as digestor slurry, in common with other organic wastes, supports fish growth by stimulating the natural food web rather than as a direct feed source, it gives less predictable yields that can be obtained using complete pelleted feeds.

Delmendo (1980) reported that in China, a 10 m^3 capacity digestor (to supply a family of four to six) produced approximately 10 m^3 of sludge and 14 m^3 of effluent per year. The slurry effluent was used in fish nursery ponds as well as for the culture of market fish. The sludge is more generally used for land crops.

Information on the nutrient content of digestive slurries is confusing. Barnett *et al.* (1978) state that anaerobic digestion reduces waste solids, concentrates nutrients, and conserves nitrogen through changing its form.

Many authors, quoted by Barnett *et al.* have found an increase in ammonia nitrogen of about 10-25% with a corresponding reduction in organic nitrogen. Coe and Turk, however, reported increases in crude protein (100%) and amino acid content (400%) after digestion of cattle wastes. Barnett *et al.*

Box 9.4. Loading a continual flow digestor with pig manure.

An average fattening pig (~ 55 kg) produces around 1·5 kg of faeces and 3 kg of urine/day.

If faeces are removed by scraping and urine is not used the waste must be diluted (1:3, faeces:water) before addition to digestor.

If faeces and urine are collected i.e. 1·5 + 3 kg = ~4·5 l, dilute using 2 l of water, digest wastes for 30 days.

Therefore, volume of digestor must be (4·5 + 2) × 30 = ~200 l/pig i.e. a digestor of 2 m^3 capacity (minimum) is required for 10 fattening pigs.

For a pond system of 1 ha and 100 pigs, a 20 m^3 capacity digestor should be adequate.

Systems using water to clean pens should be of larger volume, but the dilution of wastes by this water should not exceed 25 l/pig/day.

(1978) suggest that the amount of the more available ammonia depends on the type of materials digested. Thus, the amount of ammonia present increased when sugar cane bagasse or high protein leaves were mixed with manure. It might be expected that organic nitrogen, in the form of protein contained in the rapidly increasing number of bacteria and other microbes, might be greater during and after digestion than in many raw wastes. However, with manures and other wastes originally containing large amounts of bacteria, e.g. human faeces, the proportion of organic nitrogen may fall.

A loss of approximately 20-30% of the organic matter occurs during digestion (Barnett *et al.*, 1978); this might be expected to reduce food available in the pond via heterotrophic pathways when compared with that obtained from fresh waste addition. Indeed, in some respects digestion replicates the heterotrophic pathway in pond productivity, breaking down organic materials and producing elemental nutrients. On balance the high quality microbial protein together with an immediate and rapid fertilisation effect of a slurry rich in readily available nutrients probably compensate for any loss of carbon. In wastes that are relatively high in carbon and low in nitrogen and phosphorus such as some plant materials and ruminant faeces, digestion may provide the pond with a better balanced feed/fertiliser.

(b) Dried slurry and sludge

Replacement of dried digested slurry for concentrate feed was found to depress yields of the blue tilapia (*O. aureus*) (Degani *et al.*, 1982). This is not surprising if the reduction in nutrients in sun dried slurry is considered. In Israel, digestor sludge (total fresh nitrogen ~2·16%) is often dried prior to use in agriculture. Drying looses an estimated 18% of the total nitrogen, mainly in the volatile ammonia form.

(c) Biogas digestors

The rapid expansion of anaerobic digestion in some developing countries as a means for providing cheap energy, has not met universal success. Crocker (1983) suggests that lack of institutional support, in the form of a network of trained and motivated technicians, is often the primary reason rather than any lack of capital. In the Philippines and Thailand for instance, with the exception of agro-industrial plants, only 15% of digestors were working satisfactorily. The capital cost, especially of designs necessitating steel gas holders, is also a limiting factor in many cases.

There is no doubt that, on an agro-industrial scale, continuous anaerobic digestion can be a most effective and flexible means of extracting value from slurry waste, whilst preventing pollution. Anaerobic digestion may be more flexible for the treatment of nightsoil slurry on a municipal scale than composting which would require large quantities of bulking agents and labour. Commercial production of fish on a level with that using conventional pelleted feed would require large amounts of slurry (~1·5 t/ha/day) and this may raise questions as to the capital cost and digestor design for these amounts; certainly the value of the gas would have to be worthwhile. The wider use of vegetable/plant waste and residues for digestor feedstock has been suggested by many authors, since their gross volume has been conservatively estimated at a thousand times as great as that of livestock waste (Barnett *et al.*, 1978). On an agro-industrial scale, batch load digestion has been shown to be a possibility and this is undoubtedly an area of potential. Smaller scale continuous designs of digestor are more suited to slurry digestion.

In addition, methane production from liquid processing wastes has considerable potential. Packed bed or random medium reactors, which contain large numbers of balls or hollow tubes to increase surface area, can be used for

Box 9.5. Loading a batch digestor with chopped water hyacinth.

Daily fish pond loading requirement of fresh, chopped water hyacinth for optimum fish yield = 150 kg COD/ha/day or on the basis of fresh weight = 3416 kg/ha/day.

In digester, assume ~ 40% of COD lost in 15 days fermentation time

so $\frac{3416}{60} \times 100 = 5693$ kg/ha/day required for loading the digestor, which is ~5693 l.

The normal loading of batch digestors is ~7-8% solids (= chopped water hyacinth), by volume.

Thus for a 15 day retention time, volume required to handle the equivalent of 5693 l/day is $5{\cdot}7 \times 15 = 85{\cdot}5$ m^3.

As the pond must be fed at least once every 5 days, we would require 3 digestors each of $85{\cdot}5/3 = 28{\cdot}5$ m^3 operating sequentially.

such wastes. Retention times can be very low and reaction rates high, since the high numbers of bacteria are maintained in the digestor rather than being washed out with the effluent. In the longer term, genetic engineering for improved strains of methane producing bacteria may hold promise for further benefits in the areas of process stability and gas yield. Primary anaerobic digestion as a pre-treatment for rubber or oil palm processing wastes, for instance, may appreciably improve the performance of integration with fish culture.

9.3.5 Silage

General considerations

Silage production is a process for preserving and storing wet residues, traditionally green plant material. Preservation is achieved either by adding acid or by stimulating the anaerobic production of lactic acid by bacteria. Both these processes lower the pH to prevent microbial spoilage. The manufacture of silage from trash fish or fish processing wastes is a cheap way of storing a high protein feed for fish or livestock.

Silage may be produced using low technology means, and can enabl advantage to be taken of crop residues and surpluses. Raa and Gildberg (198 have summarised silage production processes:

— by bacterial fermentation (= fermented silage); adding fermentable sugars which favours a growth of lactic acid. The lactic acid bacteria produce acid and antibiotics which destroy spoilage bacteria
— by adding acids, inorganic and/or organic (= acid preserved silage).

Only bacterial fermentation will be considered since this form of preservation does not depend on acids, of which the handling and economics of use may be problematic on a small scale.

Fish silage

The use of additional acids in industrial fish silage production originated in Scandinavia. A thorough review of this technique is given by Raa and Gildberg (1982), who identified preservation by fermentation methods as having greatest potential in developing countries. Fish wastes, as a product of processing marketable fish, undersize or undesirable species, may form a considerable part of yields from fisheries or farms.

The fact that only small quantities of fish wastes may be available, or their supply irregular or geographically dispersed, makes their use problematic. The rapid decomposition of fish products in general often leads to their waste or under-utilisation; silage production may allow local production of higher quality pellet feeds, and possibly the culture of high value carnivorous species of fish. Integration of marine or freshwater fisheries with aquaculture may thus be given impetus by the cheap preservation of fish wastes. This may have a special relevance to landless fishermen who have access to public waters for cage culture of fish (see Fig. 9.6).

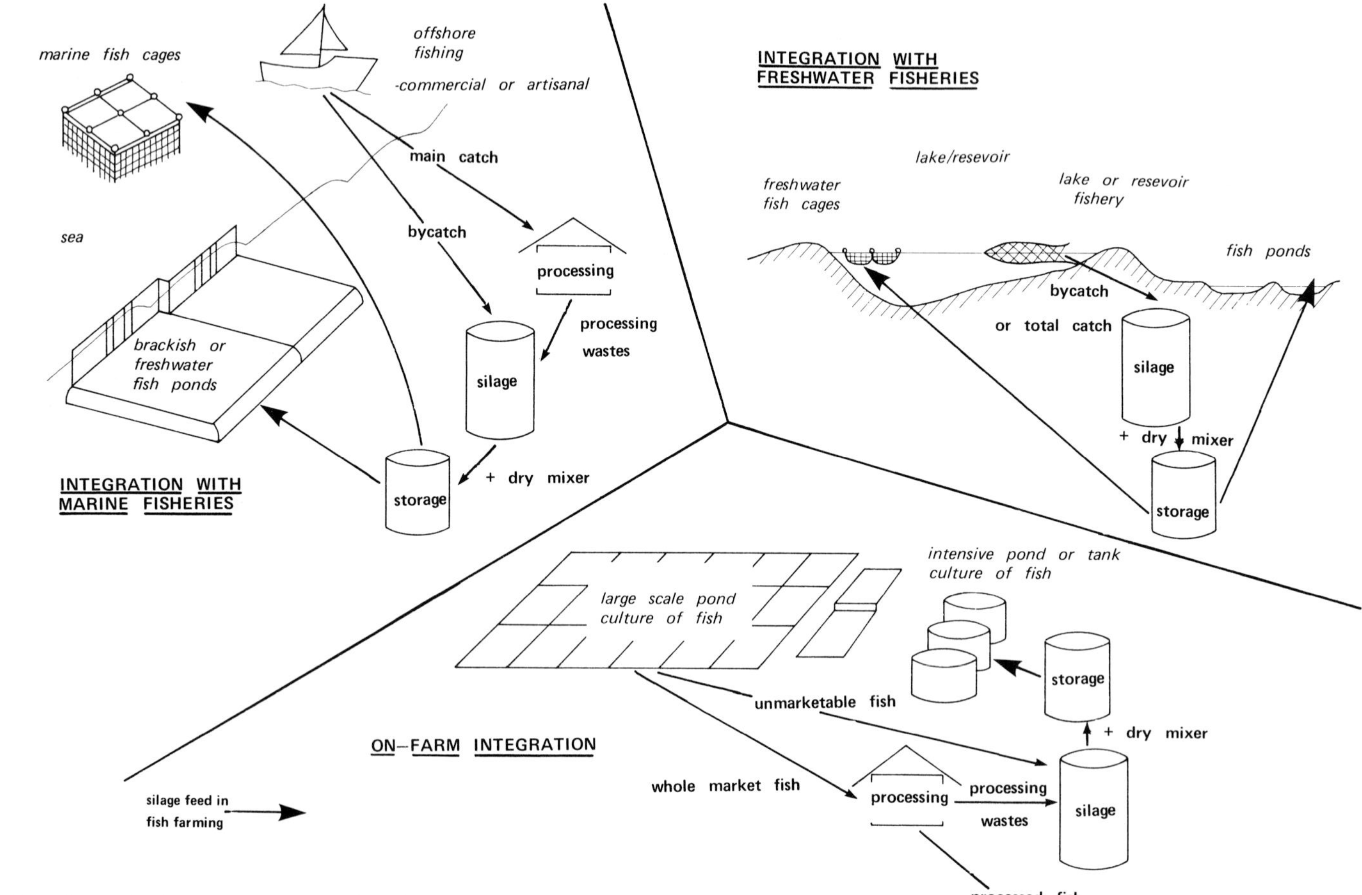

Fig. 9.6. Scenarios for integrated use and production of fish silage in aquaculture.

The integrated use of fish wastes generated on farms may also be encouraged given the flexibility of use of fresh fish silage. Farm Bikal in Hungary, for instance, produces and processes Chinese carp grown on the farm. The acid preserved fish silage made from the processing residues is then fed to European catfish (*Silurus glanis*), a highly priced carnivore. A similar use might be found for undersized or sewage-fed tilapia. In Thailand carnivorous fish are usually fed marine trash fish, an input that fluctuates considerably in cost and quality. Wee *et al.* (1986) have demonstrated the potential of sewage-fed tilapia as a replacement feed for intensively raised walking catfish (*Clarias batrachus*). Ensiled fish, when produced and used locally, can compete with both trash fish and fishmeal; the latter particularly is not affordable in most low cost fish culture systems. Fish silage is preferable to trash fish on several counts, especially because of its longer shelf life.

Box 9.6. Ensiling of fish processing wastes for re-use on the farm.

Where fish are processed on the farm prior to sale, the wastes can be ensiled and re-fed to the fish.

If a farm has 5 ha of ponds, and produces 20 t/fish/yr (4 t/ha/yr), approximately 40% of the harvest is processing waste (heads and guts) and available for ensiling.

or $20 \times 0{\cdot}4 = 8$ t trash fish/year.

on a weekly basis, 160 kg/week @ 0·8 kg raw material required per litre of silage, this produces ~200 l of silage/week.

To ensile in 5 day batches, requires containers of ~400 l capacity and ~20 l molasses*/batch. After ensiling and drying (15% DM → 80% DM), 30 kg dried material/week is produced.

Mixing with other materials could give ~80–100 kg of good quality, supplementary feed (FCR ~3·1).

When given to fish in ponds this feed could produce an extra 1500 kg fish annually.

* not always used

Silage from plant materials

There is a lack of information regarding the feeding of conventional green fodder silage to fish, although it is commonly used in the feeding of ruminants and some monogastric animals. In China, water hyacinth and water lettuce are reported to be fermented anaerobically before feeding to fish (Edwards, 1985[c]). The material is first chopped to 6 cm strips and dried to 65-75% moisture before sealing with a layer (15 cm) of dry grass or soil in ditches. Silkworm, or other livestock, faeces have also been composted with water hyacinth (see section 9.3.2). The potential uses of silage for feeding herbivorous fish, in areas of surplus green plant residues and fermentable sugars (e.g. plantations), is yet to be tested elsewhere. However, the inclusion of grains, sugar rich residues or other plant materials as a source of carbohydrate and also as bulking agents is necessary for the production of fish silage.

Methodology for silage production

(a) Simple silage production (see Fig. 9.7)

The use of dried fish silage as part of a complete diet for *Oreochromis andersoni* in Zambia has been reported (K. Jauncey, personal communication). Low value, wild fish are routinely caught in local water storage dams. After mincing and mixing with molasses at 5% by volume, a starter culture (about one litre for a 36 gallon drum of silage) is added, and the mixture is fermented for two to three days. The silage is then dried onto wheat middlings using a home-made solar drier. Wee *et al.* (1986), using chopped trash Nile tilapia (*O. niloticus*) and only cassava achieved a satisfactory silage in five to seven days without the use of a starter; when a pure culture starter was used silage could be made in around three days.

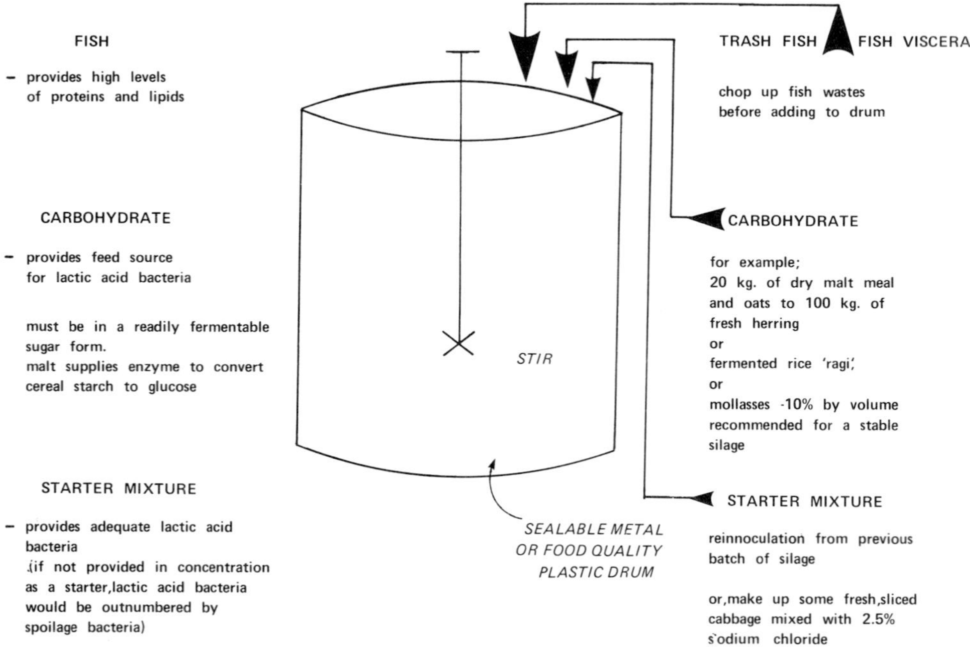

Fig. 9.7. Fish silage production by bacterial fermentation.

(b) Semi-automatic and automatic production (Fig. 9.8)

Silage production can be organised on a more automated basis if quantities involved justify the additional investment required. Equipment normally comprises simple material handling devices such as shredder pumps, conveyor pipes and tank mixers and would be expected to reduce labour requirements and improve routine quality control.

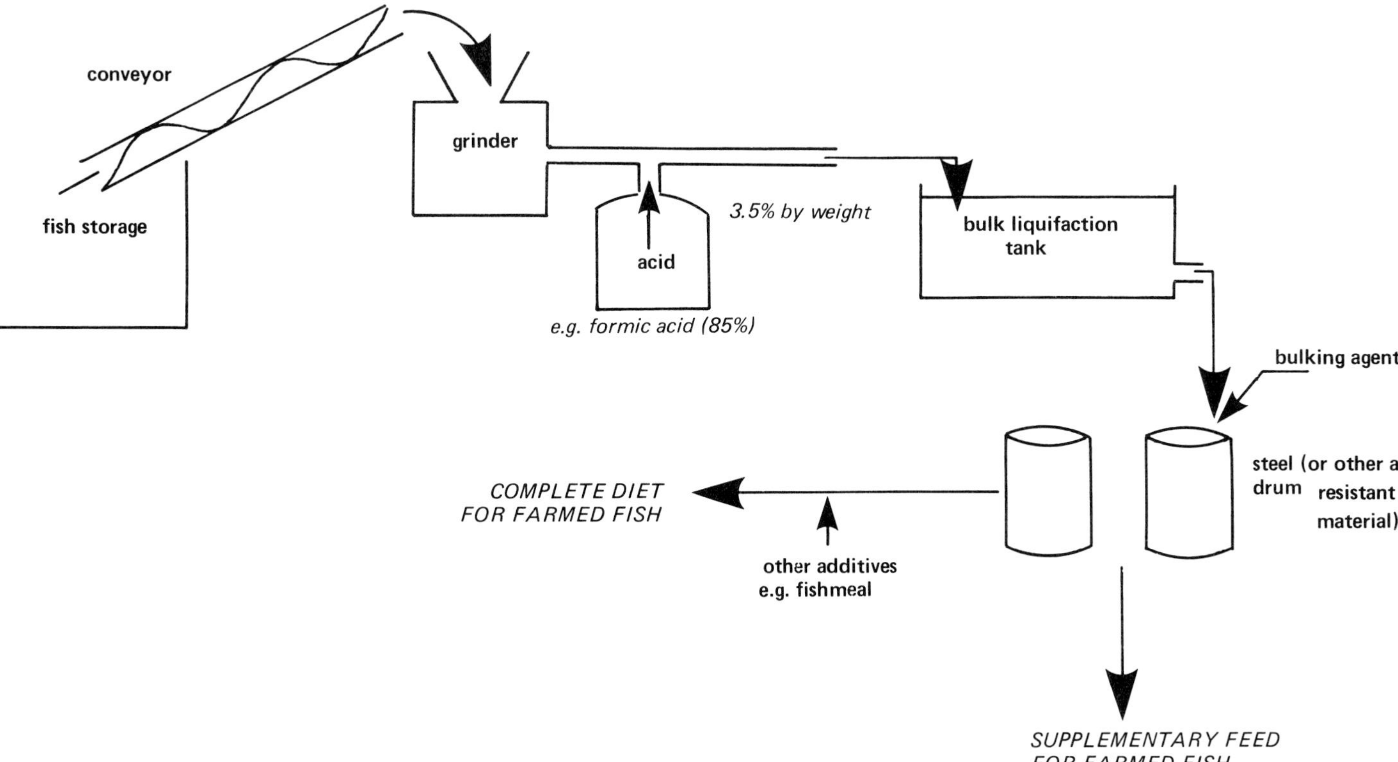

Fig. 9.8. Fish silage production by acid preservation. (After Tatterson and Windsor, 1974.)

Constraints to the use of fish silage

(a) Nutritional value
The poor nutritional value of silage in some diets has been attributed to vitamin B deficiency. An enzyme, thiaminase, contained in fish tissues, especially in carps and other freshwater fish, is known to cause this problem. Heat treatment (five minutes at 82°C) can avoid the problem but this may not be economically viable, though recent advances in heat exchangers may make this feasible.

Rancidity problems have been encountered when silage is made from oily fish (Disney and James, 1980); these can be overcome by the addition of an antioxidant (~250 ppm of BHT). Wee *et al.* (1986) found that antioxidants were unnecessary when silage was made from waste-fed tilapias low in body fat.

(b) Pathogen survival
Acid sensitive pathogens, such as the bacteria causing typhoid and cholera, are thought to be completely destroyed in silage. The survival of spores and toxins, present in the fish prior to ensiling, is not known however. Levels of virus and parasite survival are also unknown.

Other microbial and single-celled processed feeds
The growing of bacteria, fungi and algae using various wastes as their feed substrate has been widely reported as a source of processed high protein food for fish. Although the controlled production of these organisms is now technologically possible, their economic use does not look promising except perhaps on an agro-industrial scale. The potential of various single-celled organisms for fish food is reviewed by Tacon (1978, 1981).

CHAPTER 10
PROCESSED WASTES–LIVE FEED PRODUCTION

10.1 Invertebrates

The processing of wastes by feeding them to various invertebrates which can then be used as fish feeds is at present mainly experimental. The potential advantages of suitable organisms typically include a rapid generation time and high productivity. Feeding low in the food chain on organic wastes, they are nevertheless often highly palatable to many fish and provide a high quality feed (Table 10.1). Organic wastes, it has been suggested, may be a very suitable basis for pond fish production without 'processing' through invertebrates, certainly a loss of energy and nutrients will occur in this extra 'link' in the food chain. However, some fish, often highly valued carnivores, cannot thrive at high density in waste-fed ponds either because the natural feed they require is insufficient or the water quality is too poor. Concentration of organic wastes via invertebrates may be a means by which the culture of these species can be integrated into the farming system (Fig. 10.1).

Bain (1982) has identified five categories for the use of live, invertebrate feeds in aquaculture:

— as a feed for fry reluctant to take immediately to dry or pelleted feeds
— as a natural supplementary feed in an extensive culture system
— as a main dietary ingredient
— as a feed supplementing a main dry diet
— as a supplementary feed considered to have high growth promoting properties.

A scenario for the role of invertebrates in integrated aquaculture is given in Fig. 10.2.

10.2 Insects

10.2.1 Insect larvae

Silkworms
Silkworm larvae, from the mulberry silkworm (*Bombyx mori*), are traditional feeds for fish in the two major silk producing countries, China and Japan (Hickling, 1962). Fish culture and silk production have traditionally been physically integrated in these societies; silk culture providing feed inputs for

fish, and aquaculture providing fertiliser and water for the mulberry bushes grown to supply leaves for the silkworms. Silkworm larvae, as a by-product of the silk industry, are available in quantity only in these countries and in India and Thailand (FAO, 1983). They can be considered a good source of protein as well as fats (especially unsaturated fatty acids) when fed fresh. Supplementary feeding of dried and fresh silkworm, chiefly to common carp and goldfish

Table 10.1. Proximate composition of some invertebrates grown on wastes compared to fish silage and complete feeds.

Food	% Crude protein DM (N × 6·25)	% Lipid DM	Source
Soldier fly larvae (*Hermetia illucens*) (grown on chicken manure)	38–40	18–28	Bondari and Sheppard, 1981
Silkworm larvae fresh	53·9	36·2	Hickling, 1962
Silkworm larvae dried	62·1	27·2	Hickling, 1962
Silkworm larvae dried and defatted	82·8	2·0	Hickling, 1962
Termites	36·8	50·5	Bain, 1982
Earthworms			
Allolobophora longa	50·4	1·4	Tacon, *et al.*, 1983
Lumbricus terrestris	56·1	2·1	Tacon *et al.*, 1983
Eisenia foetida	58·8–64	9·0(7–10)	Tacon *et al.*, 1983
Eudrilus eugenige	60·4	12·0	Hilton, 1983
Snailmeal *Achatinia fulica*	60·9	6·1	Creswell and Kompiang, 1981
Whole snail *Achatinia fulica*	16·1	2·0	Creswell and Kompiang, 1981
Fish silage			
– molasses type, Trinidad	28·2	4·4	Gohl, 1980
– acid type, Thailand	65·0	–	Disney and James, 1980
Fish pellet complete feed	46·3	7·7	Tacon *et al.*, 1983

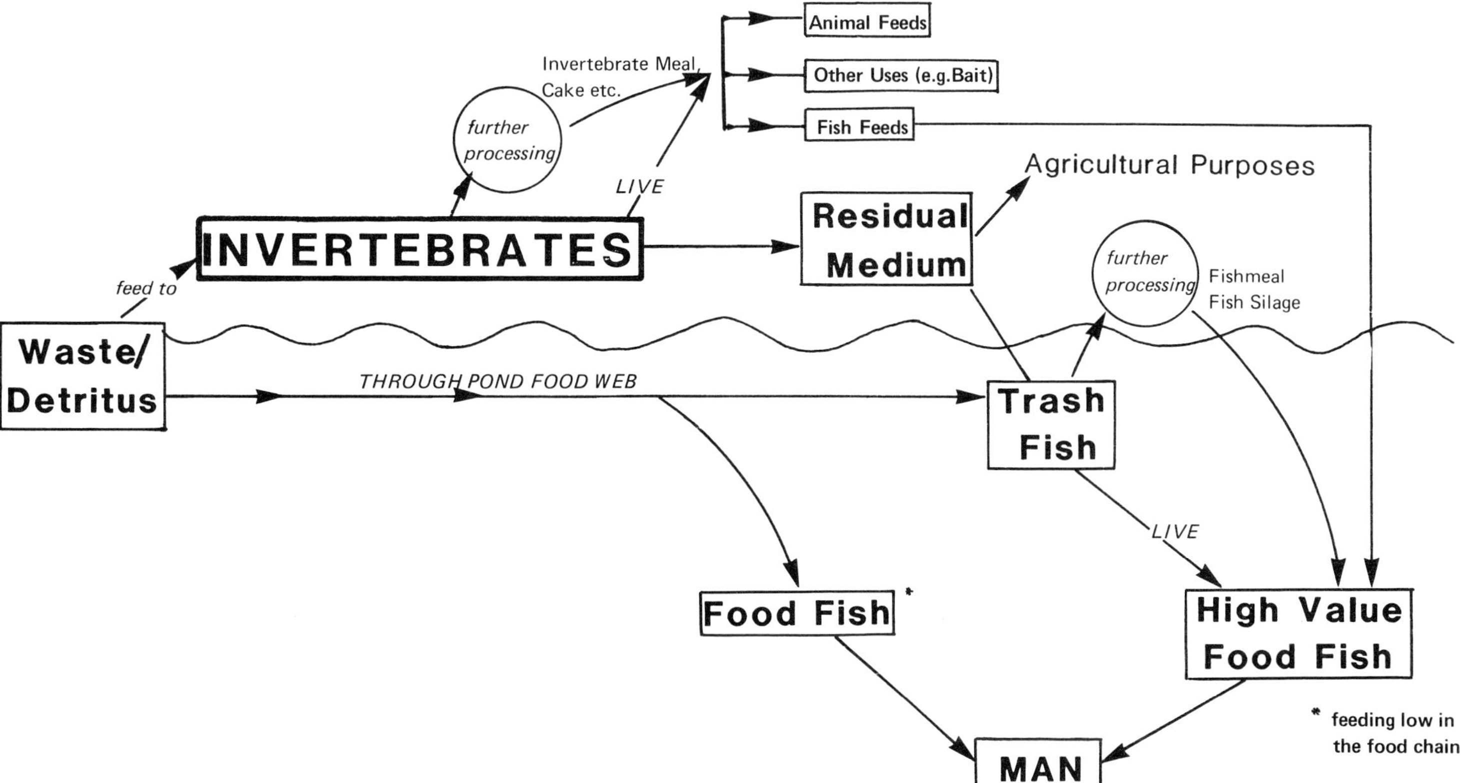

Fig. 10.1. Invertebrates concentrating organic wastes for use in fish culture.

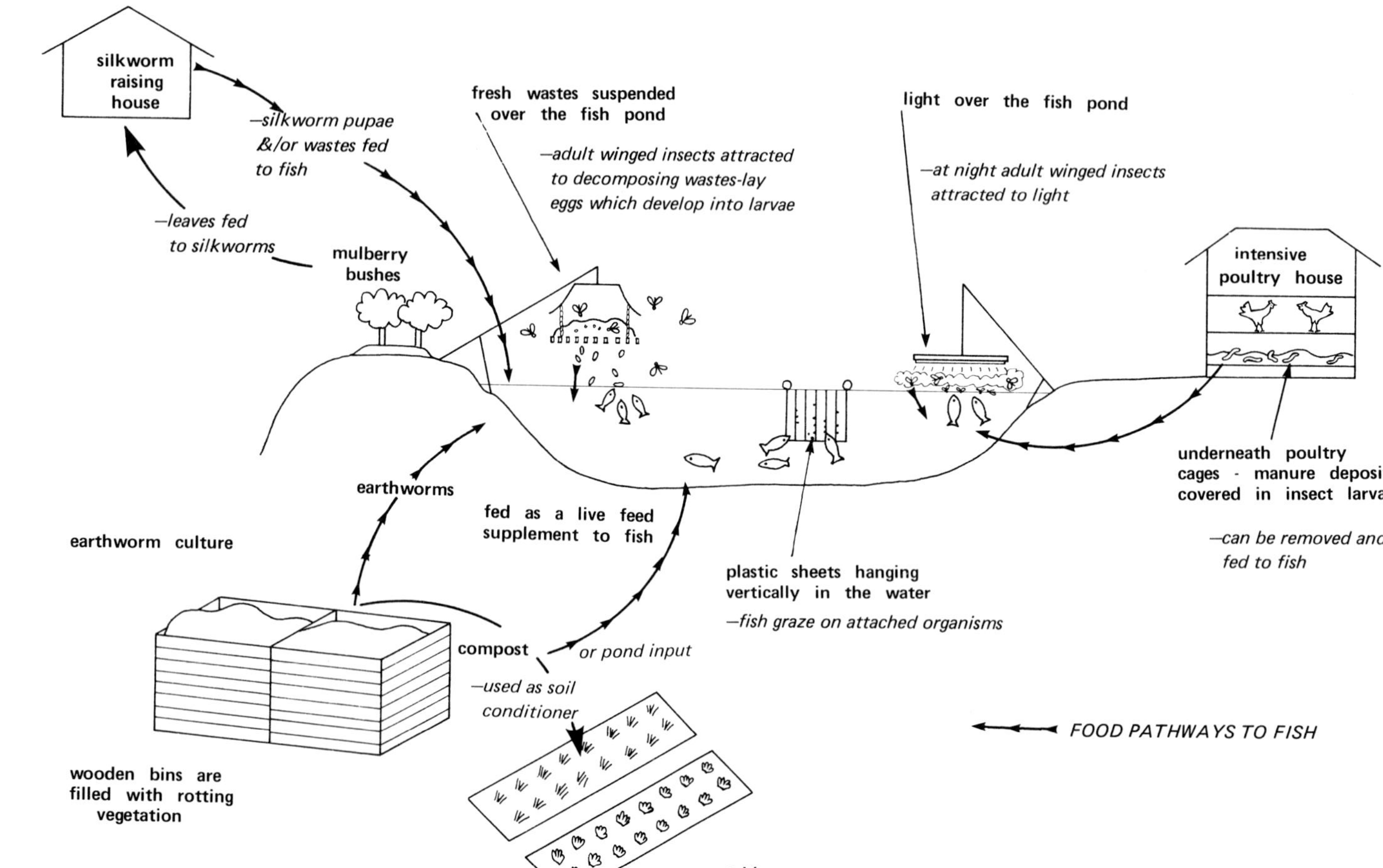

Fig. 10.2. Scenario of invertebrates as feeds for fish.

(*Carassius auratus*) in rice fields and running water ponds, has been reported in Japan (Hickling, 1962). Silkworm faeces and mulberry leaf residues are also used as fish pond inputs. In South China these inputs are an important component of integrated fish culture (section 12.4.2).

Soldier fly larvae

Bondari and Shepherd (1981) have demonstrated use of the soldier fly larvae (*Hermetia illucens*) as a feed for channel catfish (*Ictalurus punctatus*) and the blue tilapia (*Oreochromis aureus*). The fly breeds in various decaying wastes of vegetable and animal origin and is found throughout the humid tropics and sub-tropics. No significant difference in experimental fish yields were found between those fed a conventional high protein diet and a diet substituted with up to 75% chopped larvae. Further taste tests suggested that fish fed the latter diet were acceptable to the consumer. Proximate analysis revealed high crude protein and lipid levels on a dry weight basis for the soldier fly larva. These larvae have also been fed to other livestock (pigs and chickens) but an effective method of large-scale production of the larvae on waste has yet to be devised. The larvae, for experimental purposes, were grown on a solid 'mat' 2-3 cm deep over caged layer chicken manure. Extrapolations from this production suggest that up to two tonnes of larvae per month for a 20000 chicken layer house are possible. Bondari and Shepherd (1981) suggest that this figure might further be increased if better management and harvesting techniques can be developed.

Chironomid larvae

The larvae of chironomid flies (*Chironomus* spp.) are cultured for use as a live food for carnivorous fish fry in Hong Kong (Shaw and Mark, 1980). Chicken manure is used as a major input, in a culture system using converted vegetable or rice fields. The process is summarised in Fig. 10.3. Yields of 280 kg/ha/week of larvae for an input of 2928 kg/ha/week of chicken manure are obtainable. Yashouv (1970), however, obtained yields approximately ten times as great in Israel using aerated water and plastic greenhouses for shelter. Chironomid farming consumes larger amounts of chicken manure than conventional agriculture or aquaculture (Table 10.2), a factor of considerable importance where the availability of chicken manure outstrips the opportunity for use. Shaw and Mark (1980) suggest that chironomids may also be grown using piggery wastes and quote earlier work on production using sheep manure and soybean wastes. No information is available on the growth rates however. Chironomids grown naturally in fish ponds are a favourite food especially to a bottom feeding fish such as common carp (*Cyprinus carpio*). At normal fish stocking rates the levels of chironomids are low because of heavy grazing by fish (Schroeder, 1978). McLarney *et al.* (1974) suggest natural in-pond production of chironomid larvae may be improved by increasing the available substrate in the pond.

Other insect larvae

Techniques for the production of insect larvae for use in aquaculture are also being refined in China. Human wastes and decomposing animal wastes are used to attract and provide hatching sites for winged insects such as *Chrysomyra megacephala* and *Musca domestica,* the common housefly. An excavated hole of approximately 1m³, when filled with waste, can yeild 10 kg of pupae in 15

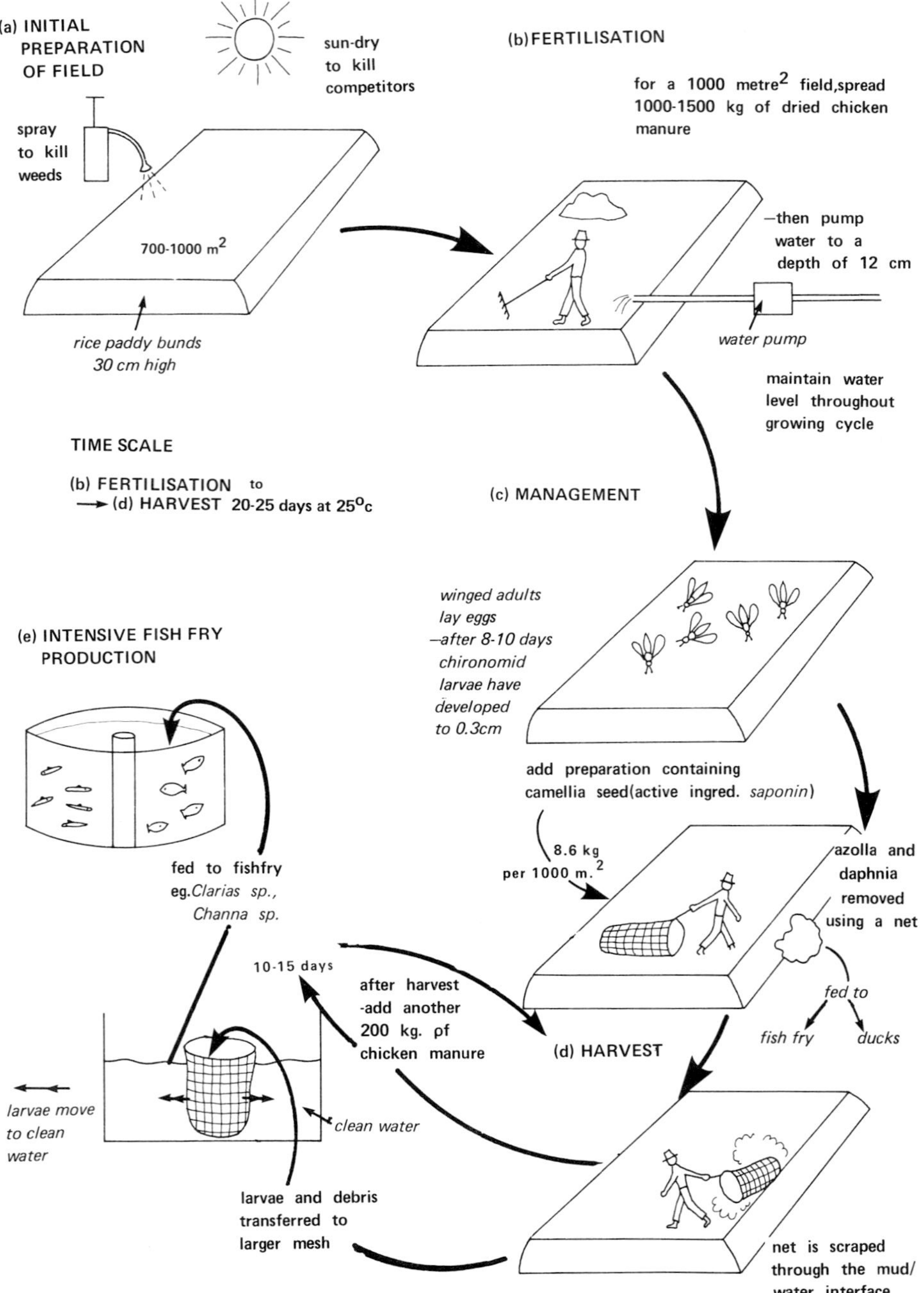

Fig. 10.3. The culture of chironomid larvae in paddy fields.

Table 10.2. Consumption of dried chicken manure by different sectors of tropical agriculture.

Used for	Quantity of manure used (kg/ha/yr)	Location	Source
Vegetable growing	21·5	Hong Kong	Lam, 1973
Integrated fish culture (*O. niloticus* – optimal manure loadings)	44·2	Philippines	Hopkins and Cruz, 1982
Chironomid larvae culture	124·4	Hong Kong	Shaw and Mark, 1980

days. The pupae are harvested manually or flushed into the pond. More elaborate box-like structures hanging over or floating on the surface of the water have also been used, designed to allow entry of the adults and the continual addition of larvae to the pond as they begin to move away from the decomposing waste (Fig. 10.2). Miller and Shaw (1969) suggested that the maggots of *Musca domestica* exhibit a photo-negative response shortly prior to pupation, and this is probably a factor in the method's efficiency. However, a conversion of manure to pupae by weight of only 30-40:1 was calculated, even under good conditions.

10.2.2 Adult insects

Termites

Termites, known for their voracious consumption of wood, have been researched mainly with a view to their control. Nevertheless, termites fulfil a highly important role in natural ecosystems and are already exploited as a food for fish in some parts of the world. Certainly, the timber associated industries generate large amounts of waste materials; termites, as one of the few organisms able to use these as food, could form a valuable link in integration of agro-forestry and fish culture at the farm level. Bain (1982) reports that two species in particular, *Reticulo termes santonesis* and *Zootermopsia nevadensis*, are easily cultured, relatively fast growing and are temperature tolerant. Further, their ability to live in 'un-structured' wood should facilitate the development of a simple method of culture. Termites, considered as a group (Bain, 1982) contain useful levels of nutrients, especially lipids, when used in fish diets, and FCR values of six have been recorded when fed to trout (*Salmo gairdneri*).

Flying adult insects

A more extensive approach to feeding insects to fish, is by attracting night-time flying adult insects to the pond surface by light. Heidinger (1971) reported that light traps using ultraviolet light significantly increased bluegill (*Lepomis macrochirus*) production, Ordinary fluorescent strip lights are regularly used over ponds in South East Asia, especially during the wet season when large quantities of aerial insects emerge. However, there is little data on the effect on fish yields or any cost benefit analysis. It might be expected that they form a considerable input of high protein food, at a time when rain-fed ponds are filling and natural food is limited. Of course, night lighting over ponds is commonly used to deter theft of fish.

Box 10.1. Growing flies on cattle slurry.

Using 1 m^3 holes in the ground, ~10 kg pupae/15 days – ~0·67 kg/m^3/day

At a conversion of 30:1 slurry:flies
0·67 × 30 – 20 l/day of slurry/m^3 hole

Thus, with 1 cow producing ~20 l slurry/day, this would suffice for 1m^3.

Would the pupae produce more fish than using the cattle slurry directly?

10.3 Snails

Aquatic molluscs often provide a major input into fish ponds in China. Harvest of wild snails has been reported to be very intensive in the Lake Taihu region, where together with aquatic macrophytes they form the basis of inputs into pond fish culture (see Table 10.3). The snails, gathered manually or by suction machines from boats, are cleaned in freshwater to remove excess mucus before crushing and feeding to fish. The black carp (*Mylopharyngodon piceus*) is a principal feeder, and when large numbers of snails are available this species may become predominant within the polyculture.

Table 10.3. Use of snails for fish culture in China (PRC).

Rate of addition to fish ponds (t/ha/yr)	Other inputs	Location	Reference
5·9	Aquatic macrophytes/green fodders	Holei Peoples Commune	Edwards, 1982[a]
100	Aquatic macrophytes/green fodders/grain	Zhang Zhuang Production Brigade	Edwards, 1982[a]
45	Organic manures, aquatic plants	Hele Peoples Commune, J'angsu.	Coche, 1980
75–112	Organic manures, aquatic plants	Other farms	Zhu de Shan, 1980

The molluscs *Stenomelania canalis* and *Melanoides* spp. have been chopped and fed as a supplemental feed to the Mossambique tilapia (*O. mossambicus*) in cages in the Philippines (Pantastico and Baldia, 1979).

Experiments using a species of terrestrial snail (*Achatinia fulica*) as feed for chickens, have indicated their feed value, especially if the shell is removed and mucus slime removed by boiling (Table 10.1). The African giant snail is widespread throughout South East Asia and the Pacific and considered a pest on vegetable and other crops. The potential may exist for the use of pest or culture snails for fish ponds.

10.4 Earthworms

Earthworms as a high protein feed for fish have stimulated considerable interest. The use of earthworms for fish bait is well established, but production to supply fish culture, particularly as a fishmeal replacement, seem to be problematic (Hilton, 1983). When used as a worm-meal, *Eisenia foetida* has been found to reduce palatability and growth response when used to replace fish meal in complete diets for rainbow trout (Tacon *et al.*, 1983). This is a worm considered most useful in waste management practices, and has been used successfully as a fish bait and for weaning juvenile eels onto artificial diets (Tacon, 1981).

Other worms that have been evaluated as fish feeds include *Perionyx excavatus* in the Philippines (Guerrero, 1981), the African night crawler (*Eudrilus eugenige*) (Hilton, 1983) and *Dendrodrilus subrubicundus* (Stafford and Tacon, 1984). The first species, when fed as a meal, was reported to successfully replace small amounts of fishmeal in diets for cage cultured Nile tilapia. The third species, when used as a freeze dried meal, could replace only 10% of fishmeal in complete diet for trout. The worms were grown on domestic sewage; at higher levels of inclusion in the diet, fish growth was reduced and heavy metals accumulated.

The use of earthworms to convert organic waste into useful soil conditioners is practised on a large scale in Italy (E. Stafford, personal communication). Their action is believed to increase the availability of nitrogen and phosphorus in the 'compost'. The use of earthworms as a fresh, live feed for cultured fish in the manner of insect larvae has received little attention. The high palatability and growth response found by Tacon *et al.* (1983) for certain earthworms (*Allolobophora longa* and *Lumbricus terrestris*) when fed fresh suggests this may be a logical route for using earthworms as fish feeds. The catfish, *Clarias macrocephalus,* was found to grow better when fed fresh slices of a pheretemoid earthworm (*Pheretima elongata*) than on pelleted commercial feed (Yakupitiyage, 1984). The worm was grown in a medium of various bedding materials loaded with human cesspool slurry. Culture methods are known for several other common earthworm species (Neuhauser *et al.*, 1980) and these might be adapted to facilitate live harvest and feeding to fish (Fig. 10.4).

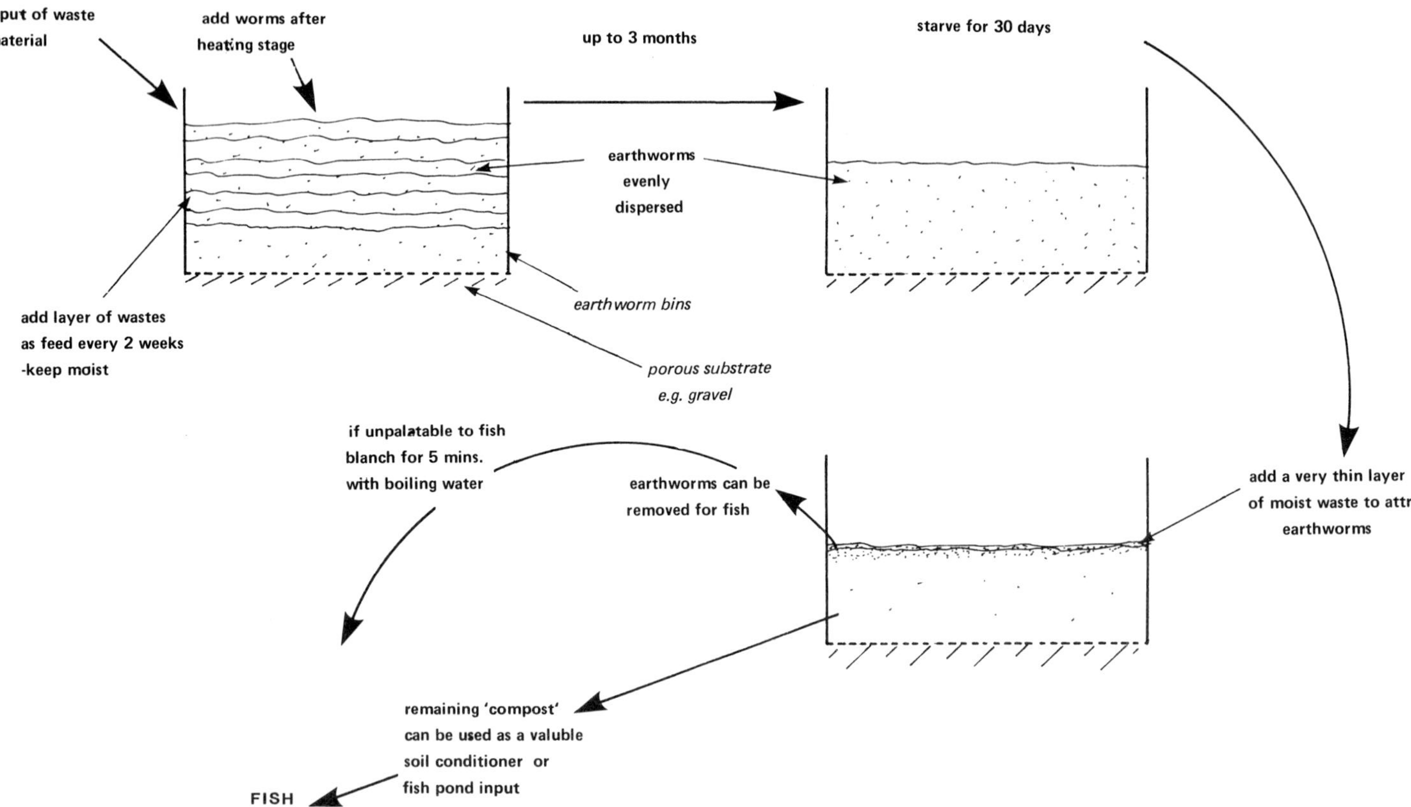

Fig. 10.4. Simple culture of earthworms for use as live fish feeds. (After Stafford, 1984.)

Box 10.2. Growing worms.

Food conversion ratio (FCR)		
	of vegetation:worms	~50:1
	worms:fish	~10:1
Therefore,	vegetation:fish	~500:1

Feeding vegetation directly to fish can obtain FCR values of 40–50. The production of worms for feeding to fish is almost certainly limited to high value carnivorous fish and/or sport fish.

CHAPTER 11
ECONOMIC AND RESOURCE ASSESSMENTS

11.1 General considerations

The profitable sharing of resources, between local centres of output whether of animal, fish or arable crop origin, is the cornerstone of integrated farming. The value ascribed to products bought from one production area will affect their subsequent use in another. Feed and fertiliser costs as a proportion of inputs into agriculture and aquaculture are usually considerable; the financial incentive to recycle by-products therefore, especially those with high food and fertiliser value, is likely to be strong. The integration of production centres to reduce the volume and cost of bought-in resources, whether food, fertiliser, water, labour, etc. may allow improvements in the efficiency of their use. In addition, social costs or financial penalties for pollution caused by waste or by-product accumulation may be avoided. It is clear therefore that direct economic benefits may be realised in many integrated systems, or that integration allows the operation of systems which may not otherwise be economically feasible.

Some of the forms in which integrated aquaculture exists have been described together with their potential yields. The production of these systems is often impressive, even when compared to intensive fish culture; but before integration is undertaken certain questions should be asked. For instance, is fish culture necessarily the best recipient of by-products and residues on the farm? Or, on the basis of cost/benefit analysis would these resources, together with the additional capital, labour and other inputs required, be better used in other ways? Does the integration of production centres, as opposed to separation, improve profitability? Fundamentally, what are the other critical areas for assessing the performance or potential of the individual production centres and integrated farm as a whole?

Some of the relevant factors which determine the success or otherwise of integrated fish culture are discussed below. The economics of substituting low value wastes for expensive feedstuffs, and the underlying factors that affect this choice, are considered. Market systems frequently change and integrated systems should be analysed for sensitivity to fluctuations. The economic interdependence of such systems may make them more or less vulnerable; market weakness causing a change in production of one component would directly affect another. For instance, a reduced market for pigs causing a downturn in their production would in turn reduce manure output and fish production. The economic basis of more complex integrated systems will obviously make analysis more difficult. Economic factors may have a different weighting depending on whether an existing system is being integrated or whether a new system is being designed. Later in this section we subject some aspects of integrated fish culture, fish-in-rice, cage farming and weed control by fish, to more detailed review.

An analysis of integrated systems should take into account more than just financial costing; social and welfare considerations may be less easy to define, but be just as important for practical success. For example, the conversion of an existing system to an integrated one will inevitably involve some investment and a possible loss of security of predictable income although potential returns are better. Aversion to such risks, for instance, has been identified as a common cause of failed aquaculture development (Panayotou, 1982). It is now necessary to consider the economic significance of some important characteristics of integrated farms.

11.2 The effect of the 'value of the resource' on integration

The resource to be integrated within a fish culture system, whether manure, a processing by-product or water, usually has a value whether or not this is apparent in money terms. The value of the same resource may vary considerably with time and place, and this will be reflected in its economic use in aquaculture.

11.2.1 Competition for high value resources

Wastes/by-products

A high nutrient content and hence capacity to be useful as a food, fertiliser or fuel gives value to by-products. Typically, wastes which are low in water and indigestible fractions are also a valuable resource, not least because they can be transported more economically than other more dilute products.

The opportunity cost—or the cost ascribed to a resource commensurate with its value if used for another purpose, is then reflected in its actual cost. Wastes or by-products with a high nutrient and low bulk characteristics will

Box 11.1. Opportunity costs.

Supposing 1 t of dried chicken manure used in the rice field will produce an extra 300 kg of rice worth $150.

If the manure is to be used in fish culture, the opportunity cost of not using it for rice should be taken into consideration.
i.e. gross opportunity cost = $150

But, if

the cost of transporting it to a distant field and spreading it as a fertiliser is	$ 20
compared to transport to and use in a nearby fish pond	$ 10
then, the net opportunity cost is	150 − (20 − 10) = $140

have higher opportunity costs than wastes with low nutrient and high bulk content. Thus, oilmeal or cakes command a higher market value than a bulky and unstable waste, such as manure slurry, which is more difficult to handle and use.

The high market value of chicken manure in the Philippines, for instance, will involve a large opportunity cost if it is to be diverted into use for fish culture. Thus, using this input to maximise fish production does not necessarily maximise profit (Hopkins and Cruz, 1982).

Processed wastes in particular, because of their high feed value, stability and more homogenous nature, are often desired by other sectors within agriculture; their competitive use in integrated aquaculture may therefore be only marginal.

Water

Other sectors of agriculture may also compete with fish culture for water. The efficiency of water use by aquaculture varies widely (Table 11.1), when the production per unit of water used is considered (kg fish/m^3 water). However, in arid lands, such as Israel, it is often the realisable profit from the use of a volume of water that is important; if aquaculture is to compete successfully for water with agriculture it must be efficient. Water may constitute 10% of fish culture operating costs in Israel (Hepher and Pruginin, 1981), and yet it is still considered subsidised in such a water-limited economy (Vardi, 1980).

Table 11.1. Water use in aquaculture.

	m^3 water/t fish
'Static' earth pond production of air breathing fish (e.g. *Clarias* catfish, Thailand)	50–100
Semi-intensive pond culture of tilapia and common carp (Israel)	500–1000
Intensive tilapia pond culture	2000–5000
Intensive cage culture	10 000–20 000
Pumped water pond culture (e.g. *Ictalurus* catfish, USA)	10 000–20 000
Intensive, flow through carp or tilapia culture	50 000–500 000

In irrigated fish-in-rice culture, water consumption may be uncompetitive when compared to rice culture alone. The water requirement, if fish are cultured, can increase by two to three fold because of the necessity for deeper water. In an irrigated area, this may result in water shortage for successive crops or crops of other farmers. Other economically beneficial practices may also be prevented by fish-in-rice culture (section 11.5.1).

Intensive cage culture of fish in public waters may also be economically incompatible. Enhanced eutrophication of such waters caused by fish culture considerably reduces its value as a potable supply, because of health considerations (Beveridge, 1984). The prospective benefits from using lakes and reservoirs for cage culture should therefore be balanced against the cost of making available alternative potable water.

11.2.2 Low value resources—limitations to their use

Wastes and by-products that are bulkier and have less intrinsic value as feed or fertiliser might be expected to attract little competition for their end use. However, their use in aquaculture may still be uneconomic especially if the production is not physically integrated with fish culture. The reliance on high quality maize silage as a fish pond input on some farms in Hungary, despite the availability of wastes from nearby pig production, illustrates this fact. Trials have shown that yields from carp polycultures using only pig slurry can be as high as those regularly obtained with the maize silage, but the logistics of transporting enough of the low value (very dilute) pig wastes over even a short distance for production scale aquaculture are uneconomic. The locally obtainable maize, together with a reduced quantity of manure, is used instead.

The economic necessity of physical integration is also suggested in China where the planting of green fodder feeds close to the fish ponds in which they are to be used is practised. In general, the lower the feed value, the nearer the source the feed must be for the use in pond fish culture. The rationale for using high bulk waste as inputs for aquaculture, on a farm in the tropics rather than for other uses is explained later (section 11.8).

Wastes sometimes have a zero or negative opportunity cost; in such cases their use in fish culture will have an economic potential limited only by yield, a reduced capacity for waste absorption or increasing costs of other inputs. Financial pressure via anti-pollution legislation may make aquaculture an attractive proposition for treating bulky agricultural wastes. Aquaculture has also been proposed as an economic system of human waste disposal, especially since the alternatives may be expensive in terms of capital and energy. Systems for utilising and treating both human sewage and 'dry' wastes for fish culture have been outlined elsewhere (section 8).

In the case of septic tank storage of wastes, it is possible the costs of sludge collection could be offset by the profits of susbsequent fish culture. Edwards (1980[b]) has proposed such a system for Bangkok, Thailand, in which if approximately half the sludge available could be collected and used, over 4000 t of Nile tilapia could be produced in earth ponds. Even allowing for a low market value as trash fish ($0·125 per kg), this might still realise a gross return of over $500000, while considerably relieving sanitation problems. A more thorough assessment of such a system in terms of return of capital, etc. compared to more conventional forms of waste disposal is currently under study.

Since optimal conditions for sewage treatment and fish growth are not identical, land, capital and labour requirements will be higher for a fish culture/human waste system, but it is thought such costs could be recovered by the overall economic viability of the system.

11.3 The economics of using wastes in aquaculture

Fertiliser and/or feeds are major costs in intensive fish culture. Rabanal and Shang (1979) have estimated, for example, that fertiliser and feed costs for intensive milkfish (*Chanos chanos*) monoculture and intensive polyculture of milkfish and shrimp (*Penaeus monodon*) are 39% and 28% respectively of the total operating costs.

The total or partial substitution of such expensive inputs with cheap or cost-free waste from associated agriculture is therefore attractive.

11.3.1 Total replacement of feeds by wastes

The high fish yields attainable with wastes as pond inputs does not necessarily mean that their use is more profitable than the use of conventional feeds Wohlfarth and Schroeder (1979) have demonstrated a means of establishing break-even costs for the use of feeds and wastes using actual data from Israel, based on simple linear programming techniques. This enables a strategy to use either feeds or wastes to be formed. Both feed and wastes are assigned arbitrary monetary units; in Example 1 the buying price of feed is fixed and the price of fish is computed, and in Example 2 the selling price of fish is fixed and that of feed computed (Box 11.2).

The practical application of this type of calculation is of course limited to situations where waste materials are fairly homogenous in quality and food conversion rates are known. The method is inappropriate when prices of feed, wastes or fish fluctuate greatly. In countries such as Israel where both high and low value fish feeds are readily available, their substitution by wastes for fish culture must be subject to a strict cost/benefit analysis. In other countries where feeds are not readily available, or the market price of fish is low, the use of feeds will be more obviously uneconomic, and waste-fed systems may be the only possibility. Wolhfarth and Schroeder (1979) have linked the spread of intensive fish culture in the Western world (Fig. 11.1) with the easy availability of alternative feeds, the relatively high market value of fish and aesthetic objections to the use of wastes. The development of integrated polyculture in countries where wastes are used as pond inputs is linked to an unavailability or high cost of grains and the low market value for fish.

11.3.2 Simultaneous use of wastes and feeds

The use of feeds and wastes need not be mutually exclusive, however. In China, Crocker (1983) found surprisingly large amounts of grains (35-40% dry matter of the feed input) were used along with green fodders and manures on some communes. Ruddle (1985[b]) also found a large variation in inputs used in an analysis of pond systems in the Zhujiang delta, China. The traditional use of human, animal and plant wastes was supplemented to varying degrees with

Box 11.2. Breakeven costs for wastes and feeds. (After Wohlfarth and Schroeder, 1979.)

Example 1: Computing the price of fish for equal profits from feeding and manuring: price of feed and manure kept constant.

	Fish grown on	
	Feed	Waste
Yield (kg/ha/day)	50	32
Food conversion ratio (FCR)	2·5	2·7 (dry matter)
Cost/kg of input	3·0	0·3 (dry matter)
Cost/kg of fish produced	7·5	0·81
Total cost/ha/day	375	26

Equal profits are attained for using feeds and wastes as inputs when—
$50x - 375 = 32x - 26$, where x = the price/kg of fish.

For this example, $x = 19{\cdot}4$ monetary units – *the price of fish* at which using feeds and wastes is equally profitable.

When fish is more than 19·4, it is more profitable to use feeds.

When fish is less than 19·4, it is more profitable to use wastes.

Example 2: Computing the price of feed for equal profits from feeding and manuring: price of fish is kept constant and price of feed assumed to be ten times the price of manure dry matter.

	Fish grown on	
	Feed	Waste
Yield (kg/ha/day)	50	32
Food conversion ratio	2·5	2·7
Weight of input required (kg)	125	86·4 (dry matter)
Price of fish/kg	20	20
Total income/ha/day	1000	640

Equal profits are attained for using feeds and wastes as inputs when—
$1000 - 125y = 640 - 8{\cdot}6y$, when y = price/kg of feed or 10 kg of manure dry matter

For this equation $y = 3{\cdot}1$ monetary units – *the price of feed* at which using feeds and wastes is equally profitable. When the feed price is more than 3·1 units, it is more profitable to use wastes. When the feed price is less than 3·1 units, it is more profitable to use feeds.

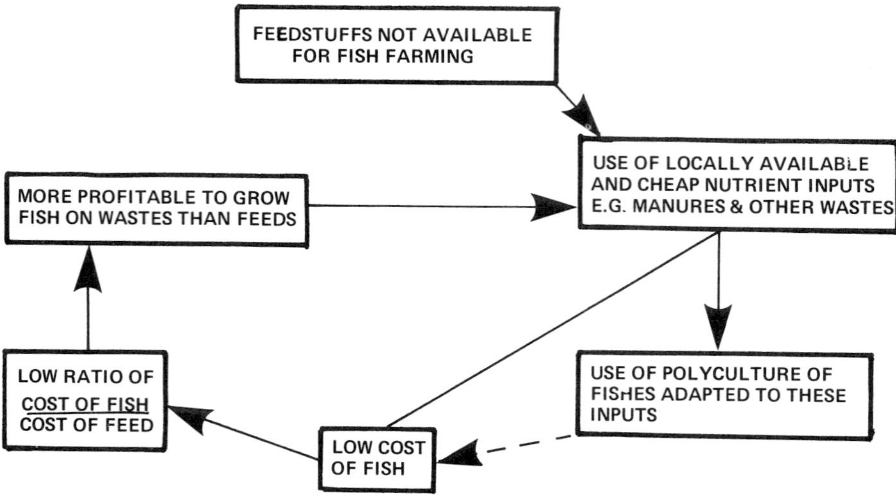

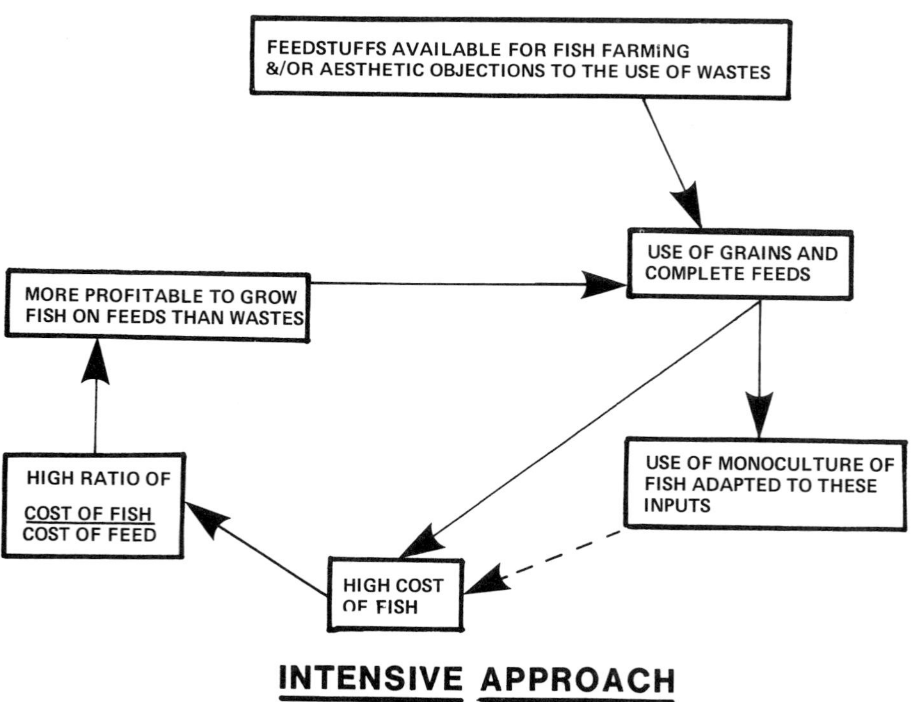

Fig. 11.1. Factors influencing the choice of inputs in fish farming.

concentrated feeds. However, farms employing less additional feeds had the lowest rate of total costs and the highest rate of return of their working capital. Profits have also been shown to improve with the simultaneous use of feeds and manure, yields of common carp increased by 24% in comparison with controls, the food conversion ratio (FCR) decreased by 0·8 units and the feed and fertiliser costs per kilogram of fish produced were reduced by 52%. Similar results were found using dried chicken manure.

11.4 Methodology for economic analysis of integrated animal/fish systems

Every actual or potential integrated fish culture system has a particular set of economic constraints determined by such factors as local land prices, market values, feed, fertiliser and labour costs, etc. (see below). Therefore, it is difficult to advocate a specific analytical method by which to assess the efficiency or potential revenue of the system. Instead, some examples are reported below which apply to specific case studies. These illustrate the types of approach required for system assessment and optimisation.

Hopkins and Cruz (1982) have analysed pig/fish, duck/fish and chicken/fish systems in the Philippines, using the following basic methodologies:

— determination of the relationship between fish yield and manure load—the *production function* (Fig. 11.2)
— calculation of the capital costs and operating costs for several pond sizes using acceptable design parameters (Table 11.2)
— determination of the relationship between pond size and capital and operating costs for excavated and levee ponds with either gravity or pumped water systems, using regression technique
— determination of the combination of pond size and manure load which maximises operating profit (five step, trial and error, computational method)
— comparison of animal density (or manure load) which would maximise the *internal rate of return* (IRR),

$$\text{where IRR} = \frac{\text{total revenue} - \text{total costs}}{\text{total capital costs} + \text{average working capital}}$$

On the basis of these methods a comparative budget for a number of animal/fish integrated operations was determined, including analyses of profit and IRR. Since the economics of integrated farming systems are highly location and time specific, their methodology rather than their particular results are interesting.

Analysis of the various production centres, for IRR and operating profits, should be made for separated and integrated ventures. The calculated internal rate of return is a measure of the system's profitability and if greater than the current commercial rates of interest, indicates a potential for investment.

Thus, if the IRR for a proposed integrated system could be shown to be better than that of animal production by itself and better than commercial rates of interest, the investment would be worthwhile.

In this case Hopkins and Cruz (1982) found that the IRR of duck/fish systems was reduced in the Philippines compared to the IRR obtainable from investment in ducks alone. This was explained by the profitable nature of duck egg production compared to fish. Integration with fish culture was still a worthwhile investment however, because:

— expansion of duck production was limited by lack of available markets, and
— the IRR of fish culture was considerably higher than the commercial bank returns.

In the case of the chicken/fish integration in the Philippines, a calculated IRR was higher for large ponds (Hopkins and Cruz, 1982), although management considerations limited the pond to 2-3 ha in size.

For small scale farmers available capital for development of integrated systems may be a serious constraint. Both intensive animal production facilities and large, machine-dug fish ponds would require considerable capital; development of low capital systems for poor farmers, however, is receiving attention in several quarters.

Table 11.2. Categories of capital and operating costs for an excavated fish pond. (Source: Hopkins *et al.*, 1980.)

Capital costs:	Operating costs:
(a) Land clearing	(k) Land rent
(b) Dikes	(l) Irrigation fee (per hectare per year)
(c) Drain pipe	(m) Fingerlings
(d) Water inlet structure	(n) Labour
(e) Storage building	(o) Poison
(f) Engineering fee (% of (a) to (e))	(p) Fuel
(g) Pump	(q) Maintenance
(h) Buckets	(r) Equipment depreciation
(i) Seine	
(j) Wheelbarrow	

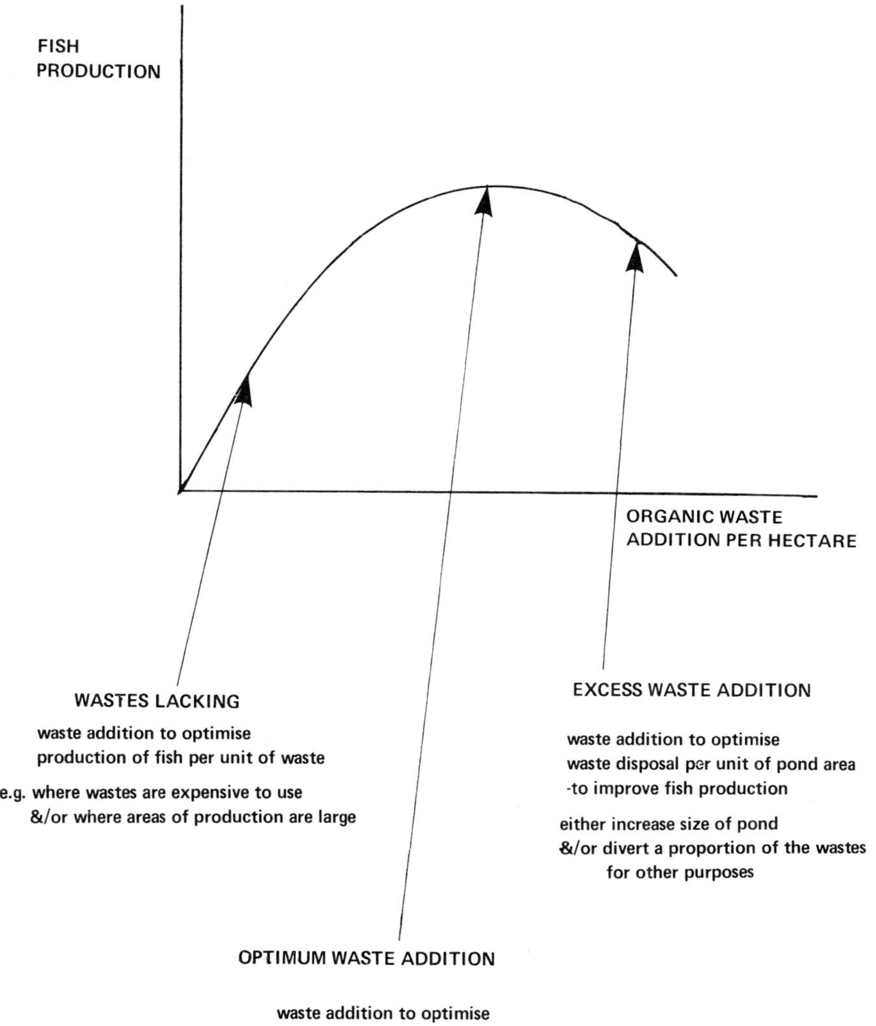

Fig. 11.2. Relationship between fish production and organic waste addition.

11.5 Integrated use of land and water in aquaculture

11.5.1 Fish-in-rice (see section 6)

Captural

Economically the importance of wild fish caught in rice fields has been considerable in those areas of the world where rice is the staple crop. In South East Asia where rice farmers typically derived a relatively low income from rice alone (Khoo and Tan, 1980), fish provided a considerable supplement. Hickling (1962) estimated that the income derived from fish harvested under the system varied between 20-33% of that from rice culture. Tenant farmers, or those with small land holdings, relied particularly on the returns from fish in paddy fields. In Java, some such farmers ceded all the rice crop to the landowners in exchange for the fish crop (Khoo and Tan, 1980). In the central region of Thailand, income from fish was found to be equal, or in excess of that generated by the rice (Pongsuwana, 1963). Since then, the gradual intensification of rice culture has reduced yields of wild fish in paddy fields and their importance in this region. Undoubtedly, this has had consequences for the income and health of tenant farmers. Where rice culture remains extensive, however, captural fisheries are usually still important. Under these circumstances rice production itself may be only a marginal or subsistence occupation; the traditional fish harvest can be critical to its continuance as a viable farming system.

Cultural

Many factors can affect the viability of fish-in-rice culture, the objective of which is to improve the return in terms of (a) unit area of land, (b) unit volume of water or (c) unit of labour expended. Many factors can affect the viability of fish-in-rice culture. Undoubtedly, the low capital requirements for fish production by this method is one of the major attractions. Middendorp and Verreth (1986) reported that costs for rice field improvement for fish culture were less than 10% of those for a pond of equivalent area. The availability of irrigation is an important factor; it has consequences for rice, fish and combined rice/fish culture (Fig. 11.3). In addition the method of fish culture (i.e. rotational, concurrent) will affect the economics of a particular system.

The economic basis of fish-in-rice culture can be assessed from the viewpoints:

— cost/benefit to the rice
— cost/benefit to the fish

The relative cost/benefit of pond culture compared with paddy culture of fish should also be considered. In a survey of farms raising fish in ponds and paddies in North East Thailand, Middendorp and Verreth (1986) found that fish yields and returns from intensively managed fish-in-rice operations could be as good as from ponds managed in the traditional, extensive manner.

The benefits, in terms of yield improvement, have already been discussed as have the possible economic disadvantages in terms of increased use and cost of water. Concurrent culture of fish-in-rice using predominantly common carp and local varieties of rice, was found to produce a healthy rate of return as a percentage of costs in Indonesia (Djajadiredja *et al.*, 1980). In all the farms analysed, fish culture, when costed separately, showed a higher rate of return on investment than rice.

In the Philippines, where rotational rice/fish culture is practised, De la Cruz (1980) has reported almost identical cost/benefit ratios for rice and fish culture. A favourable margin for rice can only be maintained in good growing seasons; under bad weather conditions fish gave more favourable results.

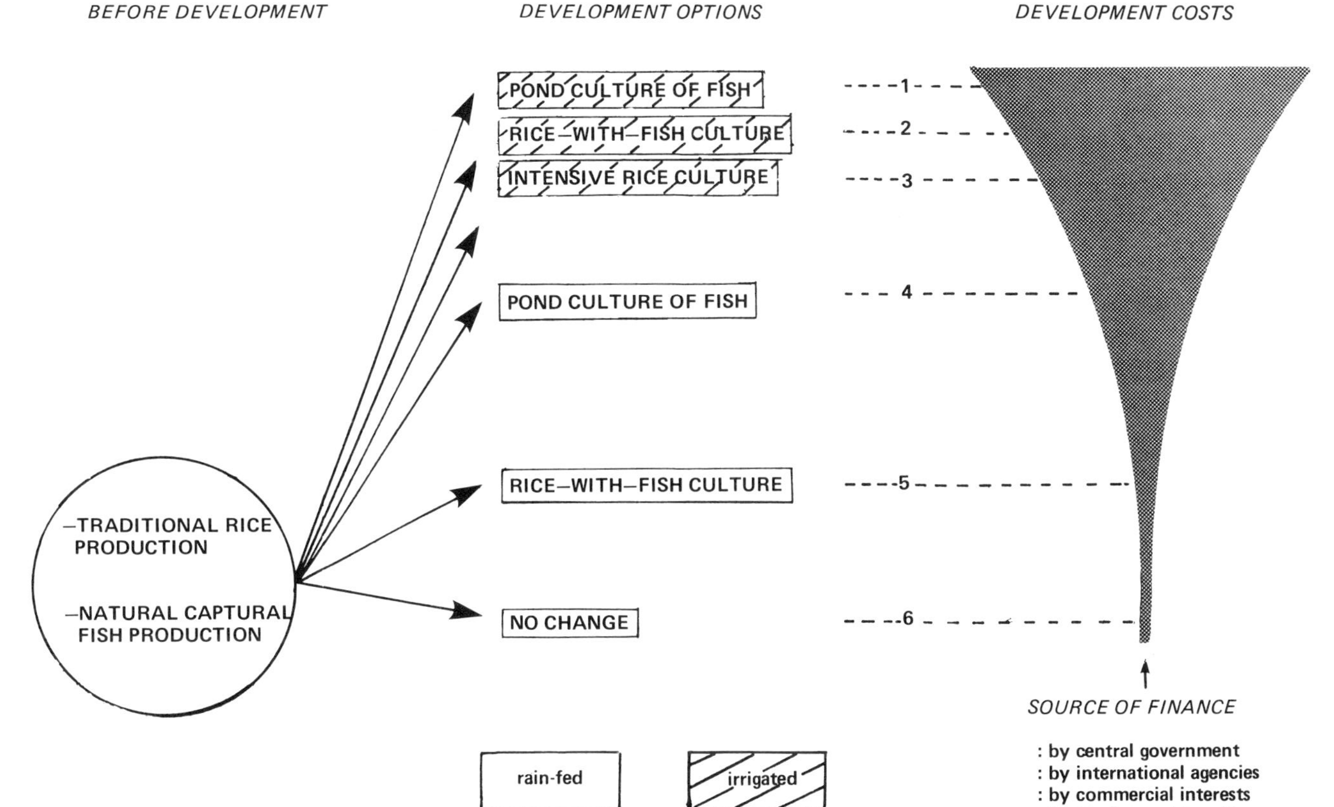

Fig. 11.3. Aquaculture development options for land under seasonal rice production.

The cropping of rice following fish culture was also found to require less fertilisers than conventional monocropping practice. This was no doubt because of the release and recycling of nutrients via the action of the fish. Although conventional fertiliser treatments gave slightly higher rice yields, the value of the yields did not compensate for the extra fertiliser costs. In contrast, Chen (1954) has determined that for concurrent fish/rice culture more fertiliser is required than for rice alone; 50-100% more than normal field practice. Taylor *et al.* (1986) showed rice yields increased by over 14% compared to controls, when fish were released in rice fields cultivated with the border-effect method (one row in four is left unplanted). Gross income also improved whilst the cost of rice seed decreased.

An additional financial advantage was found in the removal of weeds and algae when the Nile tilapia and common carp were stocked. The bottom feeding habit of the latter resulted in the clean, tilled soft paddy bottoms being

Box 11.3. For a $200 capital investment, does the farmer improve his paddy field to grow fish or dig a small fish pond?

Before investment:

well levelled rice field area	:	5000 m^2
rice production	:	1170 kg
natural fish production	:	15 kg

After investment:

	Ditch and dyke for fish-in-rice	Excavation of small pond in one corner of paddy field area
(a) Area available rice production	~4440 m^2	~4664 m^2
Rice production	(b) ~1170 kg	$\frac{4664}{5000} \times 1170 = 1091$ kg (c)
Fish production	~125 kg	45–120 kg

(a) Excavation cost ~ $0·65/$m^3$

(b) Rice production not significantly different from unadapted rice field.

(c) Depends on amount of feed/fertiliser input. Lower figure is average for Lam Pao, Thailand (Middendorp and Verreth, 1986).

(d) Profit will be affected by the value of the fish; pond culture will allow an extended harvest period and higher value.

Many other benefits from small ponds to the small-scale farmer e.g. irrigation, water supply for livestock should be considered.

immediately suitable for seeding and transplanting of rice. This would further save the cost of field preparation prior to rice production.

Fry production
Fry production in many cases may be a more attractive economic proposition than raising fish to market size. Where the growing season is short, and market demand is for large fish (250 g plus), raising of fry may be the only practical means of using fields to raise fish economically. In general, capital costs are lower since the fish are physically smaller, and large areas of deeper trenching in the paddy field are not necessary. The smaller refuges and captural sumps required will also tend to reduce construction costs. In addition, variable costs—for instance, fry and feed, will be reduced compared to those required for table fish.

A short growth period can also allow multiple rearing cycles to be managed if fry are available, for a faster return on investment. Rice culture is affected less by the beneficial and detrimental effects of larger fish. Skilled labour will be required on a more regular basis, especially if more than one species of fish is stocked. Available markets may restrict potential production and profits from fish fry culture unless further investment is made in packing and transport facilities.

11.5.2 Cage culture

Galapitage (1982) has identified some of the major economic advantages of cage culture integrated with water resources:

- cage culture can expand fish production above the maximum sustainable yield possible in the natural environment by capture fisheries
- the serious overexploitation and inefficiency inherent in common ownership of large resources can be reduced
- cage culture requires less initial capital than many pond operations. This potentially allows a relatively large number of people to undertake cage culture
- it increases employment prospects and therefore opportunity for generating additional income. This may be particularly important for landless and land-poor farmers and fishermen
- it generates a more regular income than fishing, and less subject to seasonal fluctuations. Cage culture might be expected to complement fishing as far as inputs and income supplementation
- gives a more constant and increased supply of fish. This will bring benefits to both consumers and producers.

11.5.3 Weed control by herbivorous fish (see section 5.3.1)

The costs of effective weed control by mechanical and/or chemical means in irrigation systems can be considerable. Huisman (1983) has compared the costs of grass carp (*Ctenopharyngodon idella*) control with the costs of weed clearance by conventional methods. Even allowing for some additional maintenance, as grass carp are ineffective in very shallow areas, using fish is considerably more economical. In addition to savings from weed control an income from the improved fishery is obtained (Table 11.3).

Table 11.3. Comparison of costs of weed control with grass carp and by conventional methods. (After Huisman, 1983.)

	Egyptian pounds/ha						
Type of waterway	AB	C	D	F,G	H	I	J
1 ha = x km	0·5	0·7	1·25	0·5	0·7	1·25	2·5
Grass carp stock	674	481	699	674	481	270	135
15% conventional maintenance	218	153	126	167	126	90	60
Total expenses per weed control with grass carp	892	634	825	841	607	360	195
Yield of fisheries	1200	857	480	1200	857	480	240
Net costs of carp control	+308	+223	+345	+359	+250	+120	+45
Cost of conventional weed control	1451	1020	840	1110	840	600	402
Ratio of weed control with carp to conventional methods (excluding fisheries)	0·61	0·62	0·98	0·76	0·72	0·60	0·49

AB = navigable + non-navigable main canals
C = small main canals
D = large distributor canals
F,G = navigable + non-navigable main drains
H = small main drains
I = large branch drains
J = small branch drains

11.6 Social considerations

11.6.1 Introduction

In general, the success of programmes designed to accelerate aquaculture development has been poor. Panayotou (1982) has posed two related questions concerning this matter:

— what are the factors constraining the realisation and the full potential of aquaculture?
— what are the appropriate areas for government intervention?

In some sectors, integrated aquaculture development has been pointedly more successful than others. In particular, commercial intensive livestock operations in parts of South East Asia, crop and fish integration in China and extensive cage culture in the Philippines. Undoubtedly the economic viability of strongly market orientated and high investment operations will ensure their continued development. A widespread and socially desirable development of integrated aquaculture amongst small farmers will have more constraints; some argue that attempts to promote fish culture for subsistence purposes is a redundant concept. Hepher and Pruginin (1981) claim that small ponds (100-500 m^2) to supply protein for the farming family are generally failures and that development of aquaculture should accent the larger and more professional unit. Similarly, Huisman (1984) doubts the use of 'diet improvement' as a major objective for aquaculture development planning in Africa. The failure of schemes to increase rural fish production in developing countries may spring more from a lack of knowledge of the farmer's available resources and his attitudes to fish culture and water use. Small ponds can produce useful quantities of protein, but only if they are adequately supplied with wastes and/or by-products. Larger, more commercial, ponds are not a possibility for many small-scale farmers, both because of the capital cost and the large amounts of inputs required to optimise fish production. The sale or consumption of products from small integrated farms need not be exclusive of one another; a farm producing for the market would normally have vegetable thinnings, cracked eggs and small fish available that would be consumed on-farm (Edwards *et al.*, 1986). Panayotou (1982) has isolated many of the socio-economic constraints to aquaculture in general and some of these are particularly relevant to potential integrated small-scale fish culture.

11.6.2 Insecurity of land tenure

When an individual's ownership or possession of land is in dispute, or under short-term lease, investment in aquaculture is discouraged. Panayotou (1982) quotes this problem as a factor discouraging pond improvements and efficient management of fish culture in disused mining pools in Malaysia. In North East Thailand where 51% of the land had no official ownership, fear of expropriation prevented many farmers from investing in land improvements or pond construction for fear of not being able to reap full benefit. Lack of land tenure discourages any type of improvement in aquaculture but this is particularly so where long-term capital investment is necessary or where integrated use of animal wastes in fish culture requires the moving and rebuilding of animal quarters, for instance, near to a pond site perhaps distant from the home. Such activities require considerable effort on the part of the tenant and would increase losses to the farmer if eviction follows. Government action on land reform and long-term licensing is necessary if the situation is to be corrected.

11.6.3 Multiple and common ownership

For small land owners and landless farmers, aquaculture may only be possible in waters with multiple ownership or open access. Cage culture for exploitation of inland waters, such as large impoundments, irrigation tanks, canals and rivers, has been suggested. However, protection of individual returns will require long-term licensing and/or establishment of single ownership, auctioning of property rights and the promotion of cooperatives.

11.6.4 Externalities

Externalities are the effect (either negative or positive) of one person's activities on another's; i.e. one farmer's potential fish yields and profits being reduced or improved by outside factors. The reduction in fish-in-rice yields by pesticides contained in contaminated irrigation water is one example of an externality that has reduced the viability of integrated fish culture considerably in some places.

In contrast, use of a neighbour's agricultural waste for fish pond inputs is a positive externality, i.e. a positive effect of another's action on fish yields. Government measures to control negative external effects on fish culture such as uncontrolled pesticide use in rice production should be based on the relative values of pesticide use in rice and lost potential in aquaculture. It can be seen that in many cases, integrated aquaculture provides positive externalities with regard to other activities.

11.6.5 Price of credit

Small farmers typically find credit difficult to obtain, even if the integrated aquaculture venture is capable of repaying the loan and still earning a profit. The farmer may not have sufficient collateral, especially if he does not have rights of tenure to property. Many farmers in developing countries will only have access in reality to non-institutional credit, at high interest rates. Thus, even if a 100 pig unit integrated with a one hectare fish pond is profitable at 15% interest, the much higher interest rate as available credit (often over 100% p.a.) will prevent investment. Since initial capital requirements are usually high in aquaculture ventures, alleviation of this situation by government collateral-free credit for small farmers is normally necessary. Where subsistence or market production of fish is considered particularly desirable, subsidised credit may be given.

Rapid development of integrated methods may be stimulated by the selective use of these methods. Where livestock is raised intensively already, for instance, subsidised credit for pond construction to use the waste can achieve the desired integration. Conversely, where ponds are constructed already, but receive inadequate feeds and fertilisers, credit for the encouragement of intensive livestock production may prove fruitful. In Taiwan, for example, financial support coupled with anti-pollution legislation has stimulated the adoption of biogas generators and the use of their effluent in fish culture. However, these measures should be used carefully since easy credit and grants can induce dependency and lower motivation.

11.6.6 Risk

Two major types of uncertainty have been distinguished (Panayotou, 1982):

— environmental uncertainty, which is beyond the farmer's control
— market uncertainty, which is also often not under the control of the farmer.

Attitudes towards risk are likely to differ between rich and poor farmers. Risk aversion, and unwillingness to invest in a fish pond, for instance, is likely to be greatest among poor farmers because their actual survival may be at stake. Natural risk aversion, by subsistence farmers, has a good basis since they usually have limited resources of time, labour (at specific times) and capital. In addition, for small farmers already engaged in fish farming, uncertainty may restrain the adoption of new technologies or methods.

In a recent study, Edwards *et al.* (1986) found that the livestock component within a tricommodity (fish-livestock-vegetables) small-scale integrated farm, carried the most risk and cash expenses. Thus, the consequences of increased risk to the farmer should be carefully measured before livestock production is intensified to meet the needs of fish culture. It was suggested that a suitable low-risk alternative to livestock manure might be the collection and use of large quantities of terrestrial weeds or aquatic macrophytes as feeds for herbivorous fish.

Middendorp and Verreth (1986) reported that a major limitation to expansion of pond fish culture in one part of North East Thailand was the farmers' reluctance to risk more time and land to fish culture, despite the more favourable returns than from the traditional rice crop. These workers suggested that fish-in-rice culture would be appropriate, low-risk technology for many such farmers.

Panayotou (1982) suggests three possible methods that may be used by government or agencies to reduce risks and encourage uptake of aquaculture:

— reduction of technological risks through research and extension i.e. development of relevant integrated aquaculture technologies before widespread extension to the farmer
— crop insurance
— provision of subsidised credit.

However, if widespread small-scale aquaculture is desired for disadvantaged sectors of the community, intrinsically low-risk integrated methods are required.

11.6.7 Marketing

Edwards *et al.* (1986) have identified marketing problems concerning both inputs to the farm and produce from the farm as the major constraint to development of more widespread integrated farming. The limited volume and irregular nature of production restrict the market outlets available to the small-scale farmer. These workers note that before there can be real development of integrated farming in rural areas, more effective marketing networks must become available.

11.6.8 Distributional considerations

Unequal distribution of wealth and resources are inevitable consequences of market mechanisms (Panayotou, 1982). Two major objectives of aquacultural development are often quoted, i.e. to increase fish supplies for domestic and export markets, and to provide supplementary employment, income and nutrition to the socially disadvantaged, e.g. subsistence farmers and fishermen.

The former generally relates to growth and efficiency, and the latter to distribution. Integrated fish farming is especially suited to improvement of distribution, particularly when local fish culture and consumption is required. Economies of scale do exist in aquaculture, and high marketing costs of perishable products such as fish may favour central location when development is directed towards efficient production for export.

11.7 Energy budgeting

Apart from the financial and social advantages inherent in integrated aquaculture, there may be more fundamental gains. An economic assessment of resources used in a system is made in terms of money, which reflects the present day value attached to those resources (Leach, 1975). Energy analysis, or budgeting, gives information on real, longer-term costs, and allows a valuation of processes and systems in terms of their energy relationships.

Money and energy evaluations may not agree, and often sharply disagree (Leach, 1975), especially in resource rich societies where energy is artificially cheap. Energy budgeting involves costing the amount of energy required to make productive inputs available, and attributing to them accordingly a gross energy requirement (GER) value (expressed in megajoules). The actual GER of the produced article, whether a tonne of fish or kilogram of pork is thus the summation of the input GERs.

Intensive forms of aquaculture are those reliant on large amounts of capital, concentrated feed and fuel, but highly productive in terms of labour and land. Characteristically, intensive methods use large amounts of energy per tonne of fish produced, compared to more extensive forms of aquaculture (Edwardson, 1976). Intensive animal and crop production are similarly inefficient in terms of energy use however expressed, e.g. a low production per unit energy, low ratio of energy produced to energy expended or protein produced for a given energy input. In contrast, extensive methods of aquaculture and agriculture are characterised by a low production and profit in terms of land and labour but are very efficient in terms of imported energy use (Pimental *et al.*, 1973). In a comparison of farms employing both organic wastes and concentrate feeds as fish pond inputs, Ruddle (1986[b]) found the highest conversion efficiency of input to fish in the most traditional systems (a ratio of 17:1). In systems employing both more organic wastes and concentrates, efficiency of conversion was lower.

Integration of aquaculture with either intensive or extensive agriculture has been shown to considerably improve the system's overall efficiency of energy use (Little, 1983).

Local, integrated production and consumption of fish, in a manner likely to have distributional benefits, will thus also increase energy efficiency (Fig. 11.4).

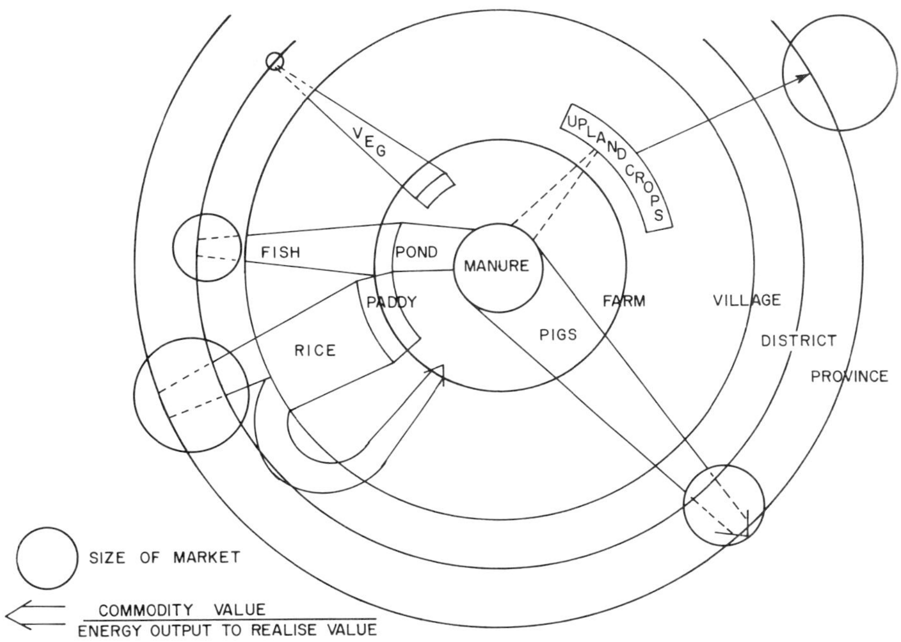

Fig. 11.4. Interrelationship of energy, distance and commodity value for North East Thailand.

11.8 Nutrient transfer efficiency

The value of agricultural wastes especially manures to the farm may be considerable. As a proportion, the costs of fertilisers compared to other inputs on the farm is often very high; thus it is incorrect to assume that these valuable nutrients should be more freely used for aquaculture rather than, for instance, field crops.

Indeed, whereas use of wastes as pond inputs may increase the energy efficiency of the system or 'conserve' nutrients, investigation of nutrient budgets shows that the level of nutrients actually extracted via the fish is only approximately 10% of the total entering. There are, on the other hand, demonstrable benefits of using organic fertilisers on the land, whether alone or in addition to chemical fertilisers, both for short-term yield improvement and for longer-term soil conservation. Moreover, whilst the use of manures for grain production may yield less protein, energy yields are significantly better than those obtained by using the same amount of manures for fish production (Fig. 11.5).

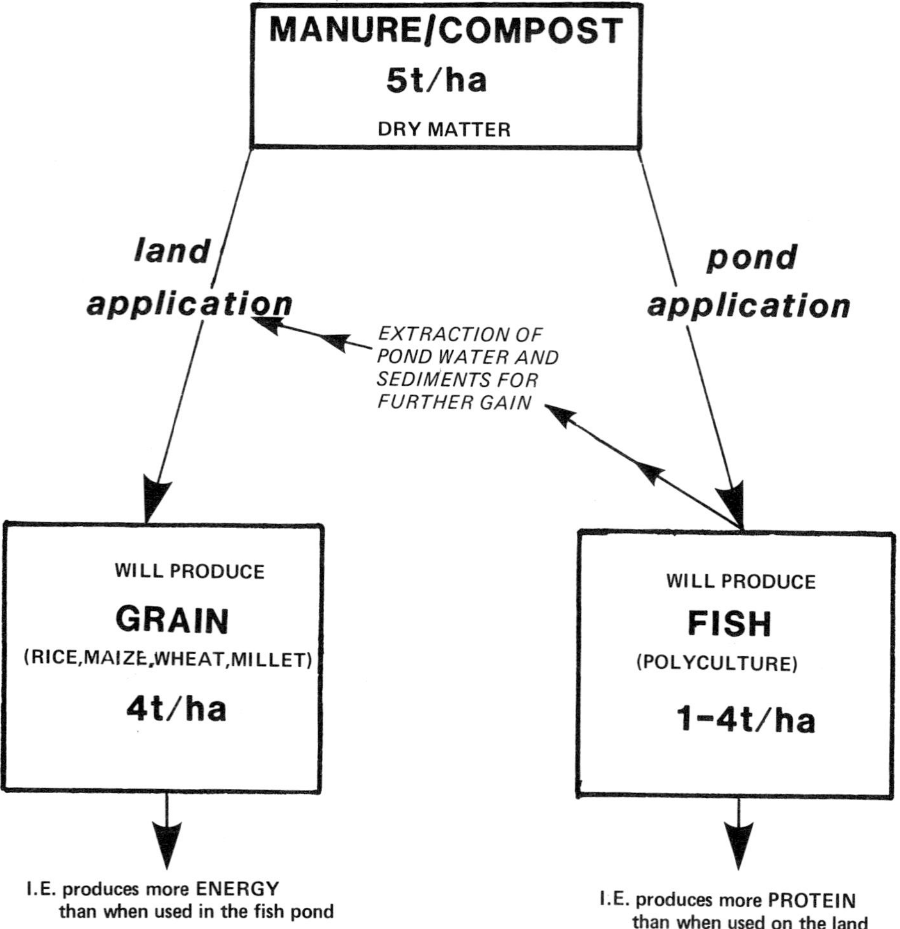

Fig. 11.5. Comparison of the use of manure as a land and pond input.

The existence of large surpluses of organic wastes in farming reported by many authors, even when chemical fertilisers are unavailable and the soil is nutrient deficient, suggests the wastes have a low opportunity cost and may be available for use in fish culture. The reason for this reluctance to use wastes has been ascribed to a lack of knowledge of their value, coupled with dislike of their handling. When used, they are often in a dry, more attractive, form although this has consequences for their nutrient content. The situation derived in Fig. 11.4 relates to a non-irrigated subsistence rice and upland cropping system in North East Thailand, but may be applicable to other farming systems. The transport and use of fresh wastes in the fields without machinery is highly labour intensive. Also, the nutrient losses inherent in the use of fresh manures under rainfed conditions are considerable (Taiganides, 1977). Disposal of bulky and unpleasant wastes into an adjacent fish pond capable of using a proportion directly and conserving the remaining nutrients for subsequent use can be an attractive alternative. Rice cropping areas would normally be closer to the source of manure, but the highly seasonal demand for fertiliser for rice production again results in high nutrient losses during storage of wastes. Limited markets for vegetables and their perishable nature often preclude the intensified use of manures for their production.

11.9 Integrated aquaculture in the development context

Box 11.4. Three unlikely integration projects.

The All American Euroafroasiatic Development Agency hereby calls for tenders on the following projects to be implemented in 1990.

(1) The promotion of integrated cattle/fish production amongst the nomadic tribes in Burkina Faso.

(2) The promotion of intensive pork production to provide the necessary inputs into integrated tilapia culture in northwest Pakistan.

(3) The planning and implementation of a carp pond system for the disposal of the wastes of the Chinese community in Singapore.

(1)
- is a nomadic life style suited to pond production of fish?
- if the tribespeople could be persuaded to stay in one place, with consequent effects on the environment, could livestock wastes be collected for use in ponds?
- would these people like fish anyway?

(2)
- is pork production in a predominantly Islamic country acceptable? Economical?
- if so, can pig waste be used as a fish pond input?
- would the people in Pakistan prefer their indiginous carps to tilapias?
- is northwest Pakistan warm enough to raise tilapia?

(3)
- Singapore is a small and almost totally urbanised country—would there be enough land and water available to process all the wastes?
- Singaporeans are generally prosperous and enjoy access to a wide variety of foodstuffs; could the carp produced be profitably marketed? (p.s. more Mcdonald's hamburgers are eaten per person in Singapore than anywhere else in the world!)

Previous sections have questioned the viability of fish culture when integrated within the farming system. An assumption that fish production is desirable should be critically analysed and very basic questions posed. Why grow fish anyway? Do people eat fish? Or prefer it to other sources of protein? Is there a preference for wild over cultured fish, or vice versa? What is the social and economic background into which cultured fish are to be introduced? Can fish be obtained cheaply from a nearby reservoir fishery or are marine fish preferred? Do potential customers prefer live fresh, and will they pay a premium for them? These questions are a sample of those that should be considered before any type of aquaculture is introduced or promoted. It can be seen that integrated fish culture, like any farming activity, does not take place in a vacuum but rather is affected by 'outside' factors and influences. It should also be remembered that there may be negative effects to integration.

Box 11.5. Negative effects of integration.

Integration can have negative effects as well as positive effects within agricultural systems. e.g. following the introduction of dry season prawn farming in rice fields in the semi-saline zone of Bangladesh.

Positive effects	Negative effects
– increased income generation	– destruction of fruit and other valuable trees*
– increased employment opportunities	reduced groundwater suitable for drinking*
– increased rice production and reduced costs	– decreased grazing area and deteriation in grazing cattle

* increased salinity effects (Source: Anon, 1985b)

CHAPTER 12
COMPLETE INTEGRATED SYSTEMS

12.1 Introduction

The book has until now described integration of fish culture with agriculture on the basis of simple relationships; the number of pigs necessary to obtain a certain yield of fish, for example. In practice, a fish pond often becomes the focus for more complex integration. In addition to the animal manure fertilising the pond, the pond water is used to irrigate the vegetables, for instance. This chapter is an attempt to assess the advantages of these more complex integrated farms, to give examples of present and potential systems and to identify some of the critical pathways for the integration of other methods of food production with fish culture.

For the purpose of this chapter, complex integrated systems are defined as those containing three components, e.g. animal/crop/fish production, and multi-component systems, which contain even more interdependent components (Figs. 12.1 and 12.2).

12.2 Rationale of complex and multi-component integrated systems

Just as simple animal/fish, crop/fish systems, etc. can improve efficiency and boost output per unit effort, more complex systems may enjoy similar benefits with greater flexibility and control of inputs, supply and markets.

In practice, however, complex systems may have management difficulties, especially if one component is optimised to the detriment of another. This has been shown in Hungary, where the production of fast growing hybrid ducks requiring intensive feeding has been developed, at the expense of former varieties that foraged to obtain some of their total nutrient intake from the pond. The relative inactivity of the fast growing concentrate-fed ducks on the ponds severely reduces the benefit to fish culture from the ducks' faeces and in turn reduces the potential for ducks obtaining feed from the pond.

A large variation in the profitability of the different components can reduce the potential benefits of integrated systems. Delmendo (1980) suggested that in the rotational use of waste-fed ponds for fish and crop production prevalent in Central Thailand (see section 12.3.3) only one product is emphasised at one time. Under-development of the other components within the system often occurs because of lack of financial resources and technical skills. The synergistic effects of the integrated systems may also not be fully appreciated. A reluctance

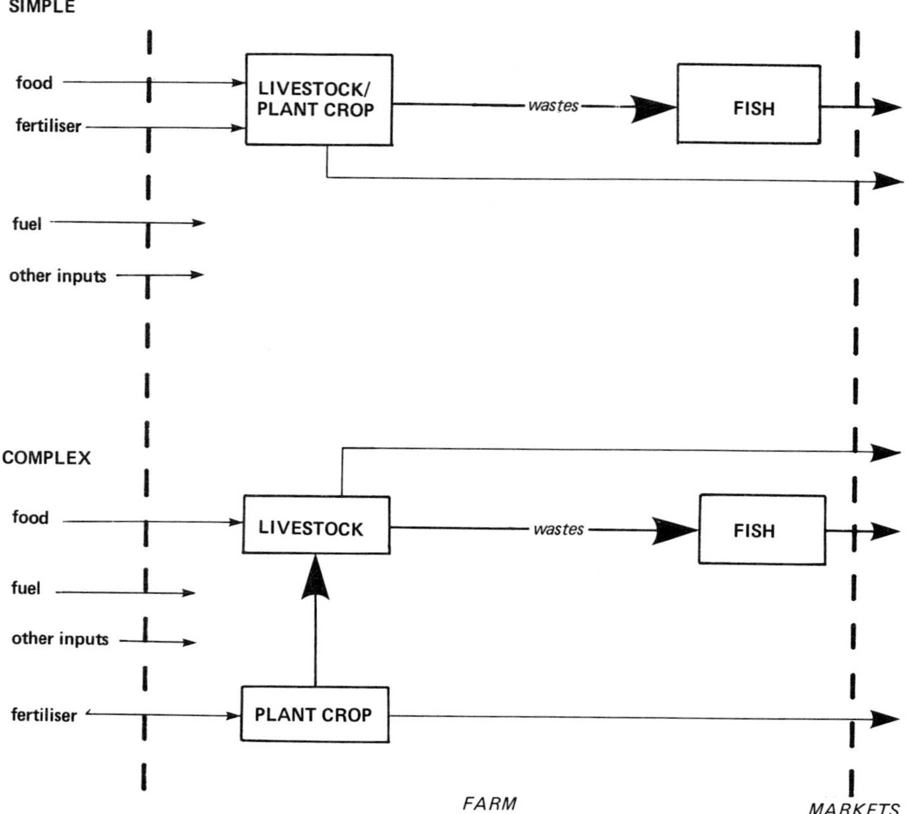

Fig. 12.1. Simple and complex integration systems.

to rotate the use of fish ponds to exploit bottom sediments, etc., if water is available, is understandable. A continual production of higher value fish, if possible, is usually necessary to optimise the return of the capital investment of pond construction and to use the wastes from animal and crop production that are continuously produced.

The management of an integrated farm requires considerable skill and knowledge. Edwards *et al.* (1986), after a detailed analysis of a tricommodity farm managed by a farmer and his family, conceded that integration must be developed step-by-step—a farmer would court disaster if he attempted to begin several different activities or 'subsystems' at the same time. Traditional, highly integrated systems, such as occur in the Zhujiang delta of China, have developed over millennia. From unpromising beginnings—the area was acid sulphate marsh of little agricultural value—the most integrated systems in the world have developed (Ruddle *et al.*, 1983).

The labour requirements for the different components of an integrated farm must be considered. In traditional farming systems labour demands tend to show seasonal peaks; the introduction of new activities will inevitably

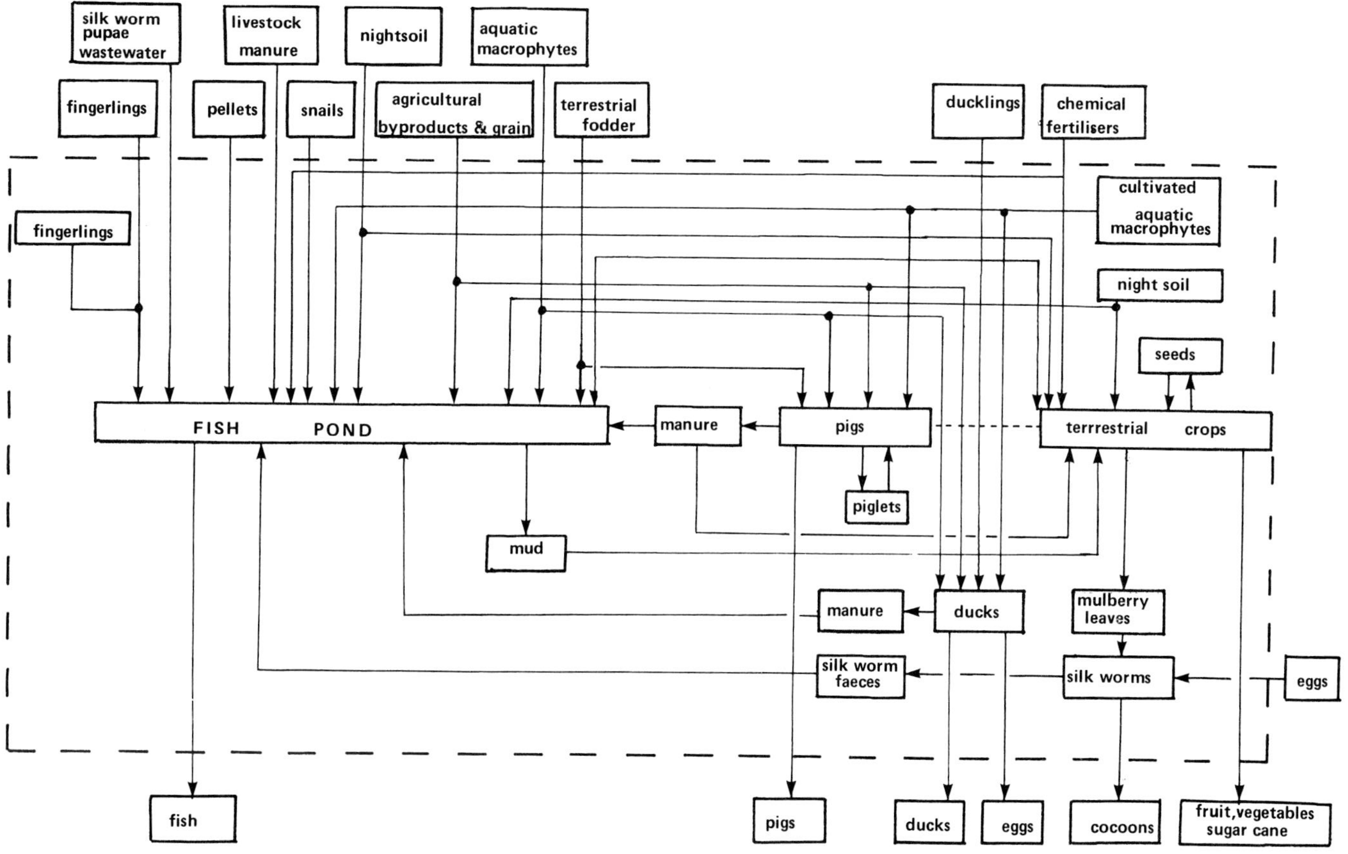

Fig. 12.2. Multi-component integrated systems. (From Edwards, 1980[b].)

change them. In a tricommodity farm in Central Thailand, Edwards *et al.* (1986) found the sequencing of labour was very different to that required for the normal rice monoculture in the area. Vegetables had a high and evenly distributed requirement. Livestock (pigs and ducks) had a low requirement, evenly distributed through the growing cycle. On a day-to-day level, the labour requirement for fish was very low; harvest and pond preparation were highly labour intensive however. Labour requirements in an integrated system in China (see section 12.4.2) showed many differences between components. Ruddle (1985[a]) found that heavy tasks (e.g. mud removal from ponds) were performed by both men and women, light tasks (e.g. picking mulberry leaves) by young people, especially girls, and skilled jobs by older, more experienced members of the community. Overall, mulberry dyke cultivation required more than 40% of the total labour effort, followed by sugar cane dykes (30%), fish ponds (15·5%) and silkworm cultivation (10·8%). In addition, although this type of system was very labour intensive, Ruddle found that it could not supply full-time employment for all the commune members; the basic elements of the dyke-pond system absorbed only slightly over half the available labour.

Complex systems can show the fish pond to be a flexible means of high protein production from a variety of wastes and other inputs. The integrated systems used in China are some of the most complex and labour intensive. The boundaries to these systems are often indistinct, and large quantities of inputs such as aquatic macrophytes and snails are obtained from neighbouring areas. In addition, an integrated exploitation of lakes, reservoirs, canal systems, etc. is undertaken on a scale not practical in most countries. Also, many of the resources used on a regular basis as fish pond inputs in China, such as water hyacinth and rice straw, are rarely used in other countries.

12.3 Categories of complex integrated systems

Complex integrated systems can be categorised by the relationship of one activity to another; for example, a farm may grow feed for animals on the farm, from which the manure is subsequently used for fish culture (plant→animal→fish), or concentrate-fed animals may provide manure for pond inputs with pond sediments subsequently used as fertiliser for vegetables (animal→fish→plant).

12.3.1 Plant→animal→fish

This form of 'vertically integrated' system is very common, but may take several forms such as:

(a) the production and use of crops and all their residues for livestock feed and then using livestock wastes for fish culture as a potentially useful way of diversifying plantation cropping. Residues and by-products from plantations can be particularly useful in fish culture if processing occurs on the farm or locally.

(b) the systems may also be adapted for extensive ruminant (or geese) production within a tree crop plantation (Fig. 12.3), or run on a smaller, more intensive scale when plantations are subdivided. Such systems could be suitable for tree crop plantations found throughout the tropics and might also involve control and use of processing wastes. Hutagalung (1981) reviews the potential for integration of tree crops with intensive and semi-intensive livestock production, and emphasises its general acceptance throughout South East Asia. Over 80% of land under plantation in Indonesia and Malaysia, for instance, supports tree crops and offers high potential for ruminant production by maximising the use of undergrowth forages and by-product feeds. End-use of livestock manures for fish culture would further diversify the rural system and improve land use.

(c) use of aquatic macrophytes grown either on the pond or in surrounding aquatic areas as feed for animals which is common in smaller scale operations. Examples of this are as follows:

— in Kenya, *Azolla pinnata* has been grown on fish ponds for feeding to chickens (N. Tudor, personal communication). A proportion of the azolla was harvested daily, dried and fed as part of a complete feed. Chicken wastes subsequently entered the fish pond, which was also fertilised with dried azolla. Thus, neither chicken production (azolla is not a perfect feed), nor fish production (azolla growing in the fish ponds would probably depress fish yeilds) is optimised, but a diversified output is obtained (Fig. 12.4).

the use of aquatic macrophytes grown or collected locally as low-cost feeds for confined livestock and the subsequent use of their manure for fish feeds is common in China and South East Asia. By-products such as rice bran, broken rice and cassava are added to improve the animals' diet but the actual composition will vary with their cost and availability. Aquatic macrophytes make variable but usually poor diets for livestock and fish alike and their use typically requires a high labour input to obtain sufficient material. In addition, when used as pig feeds, aquatic weeds have to be prepared by boiling which incurs an additional fuel cost; their use is therefore site specific (see section 4.1.1).

12.3.2 Animal → animal → fish

These systems involve the use of wastes from one group of animals, usually raised intensively, as a supplementary feed for other animals. Subsequently these animal wastes are used for fish production. The value of re-feeding animal wastes to animals has received a good deal of attention, and the method is quite widely practised at an extensive level (Muller, 1982). Caged laying chickens are sometimes housed over pigs to facilitate their waste disposal, and provide an additional source of nutrients for the pigs. The actual feed value as a proportion of that required by the pigs is likely to be small in most cases. Apart

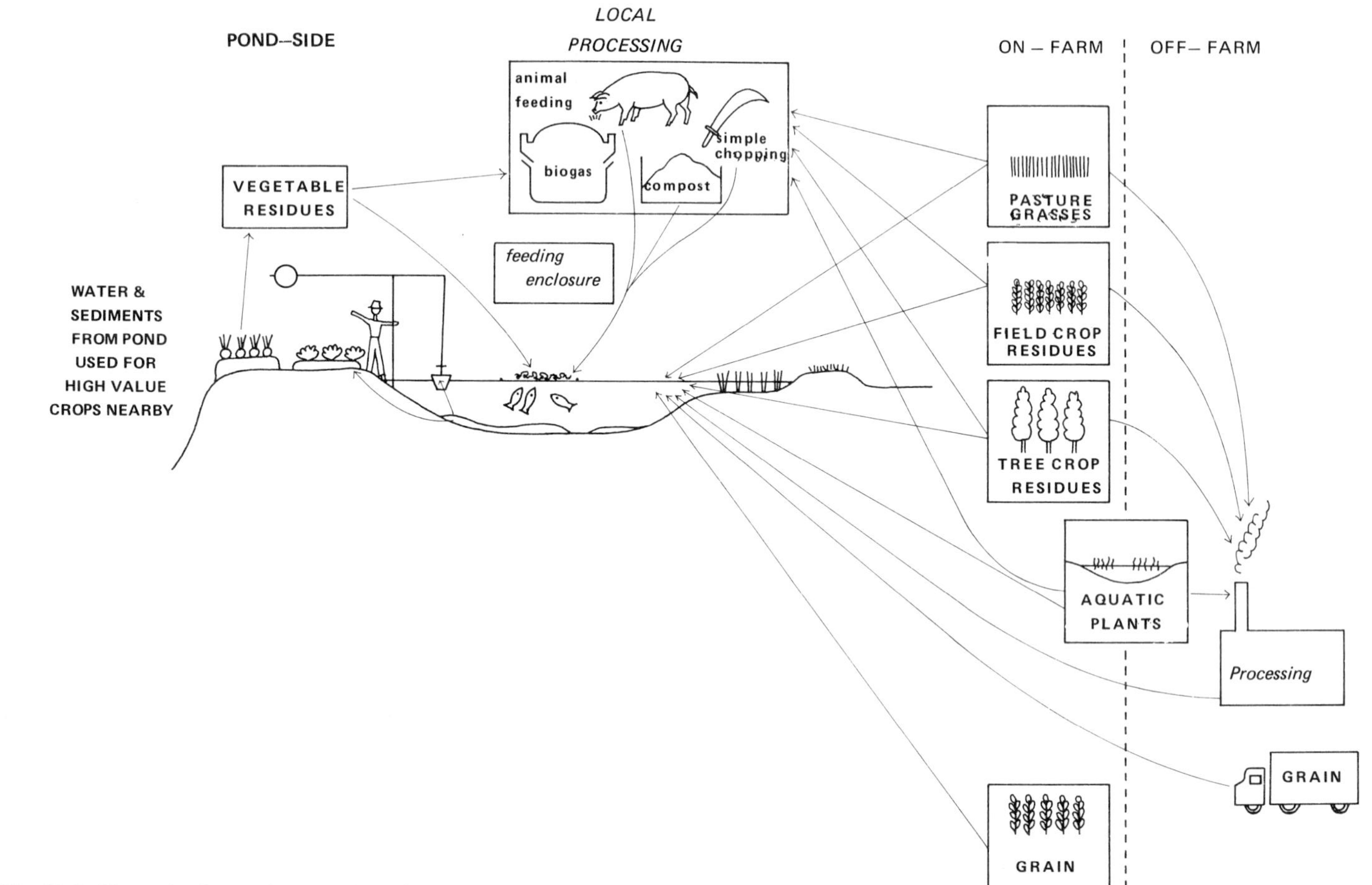

Fig. 12.3. Plantation by-products as a basis for integrated aquaculture.

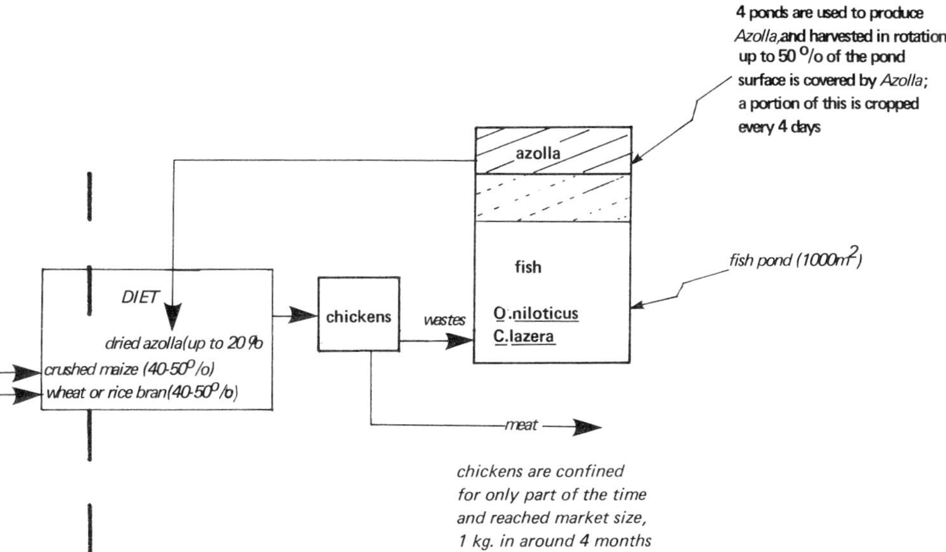

Fig. 12.4. Azolla based system for chicken and fish production.

from the supplementary food value, the practice may bring considerable reductions in the capital cost of housing and labour. Chicken manure, having the best nutrient content of the readily available animal wastes, is most widely used. Its feed value for pigs and poultry is lower than for ruminants as much of its nitrogen is in the non-protein form. Chicken manure has been fed fresh together with green fodders and waste residues, and/or in an ensiled form, to provide a complete diet for several type of ruminant under intensive conditions. The integrated development of a broiler chicken and field crop system in which wastes from both are fed to confined ruminants and their wastes subsequent fed to fish has, therefore, considerable potential.

Rabbit manure is another high value waste which holds promise for refeeding. Elemele *et al.* (1980) demonstrated its superiority to other animal manures as a dietary ingredient for broiler chickens. Since its true protein is much higher than chicken manure, it might be expected that it has potential for use as a supplementary feed for monogastric animals.

Spilt feed mixed with manures also increases in feed value when used for other animals or fish. A system of *ad libitum* feeding a rice by-product based diet to growing pigs in North East Thailand resulted in a high proportion of waste feed in the manure; this was then fed to growing ducks raised over fish ponds after ensiling to reduce pathogens.

12.3.3 Waste→fish→plant

This system involves the use of pond sediments and water from waste-fed fish culture, for further vegetable or plant cropping. Growing fish using animal or other wastes as feeds improves the overall efficiency of nutrient use within the farm system, but nutrients recovered by the fish amount to only approximately 10% of those available in the inputs. Since fish ponds conserve many of the input nutrients, mainly in the sediments, there has been interest in re-use of them via plant uptake. Several systems have been described to optimise residual nutrients following waste input into fish ponds.

Use of pond structures for growing of plant crops
(a) Small-scale
Delmendo (1980) has described a small-scale integrated farming system employing the rotation between two level plots of land, as practised in Central Thailand (see Fig. 12.5). In this system one plot is filled with water and used for manure-fed fish culture, while the other is used for vegetable cropping. The component parts vary widely depending on the interests and experience of the individual farmer. Alternate use between the two plots will allow use (by crop raising) of fertile sediments accumulated during the animal manure/fish phase.
A system employing year round input of animal wastes, combined with a seasonal fertilisation effect on rice culture, has been used in North East Thailand. The fish pond is designed to open, via concrete pipes, into a much greater area of rice fields, which can then be fertilised as required by the release of fish pond water and sediment. Thus, waste and land utilisation can be dramatically improved for fish culture during the wet season. As a bonus, in an area where paddy field fertilisation is rare and rice yields are very low, considerable improvements in rice production can result, although these are not quantitatively known. In addition, the system has potential for:

— dry season intensive cropping of vegetables in the paddy field using irrigation water from the nearby pond, and
— early rice seed bed preparations using irrigation water from the pond (see Fig. 12.6).

(b) Large-scale
A large-scale, longer term rotational system has been described for Hungary by Muller (1978) (Fig. 12.7). The system, which is only practised on a production scale experimentally, improves poor soils with little original argicultural value. The benefits are said to include a higher and more regular income, greater flexibility to changing demands of the market and improved soils. The system may have additional virtues for pond management but in general the capital cost of pond construction would be too high for adequate returns from growing essentially low-value field crops and fodders.

12.3.4 Human wastes→fish→plant

Sewage
Some large-scale controlled experimentation into using human wastes as a feed stock for an integrated fish and subsequently crop production system has been undertaken in Thailand (Edwards *et al.*, 1980). The use of high-rate stabilisation ponds for primary treatment of raw sewage, before using the algal rich effluent

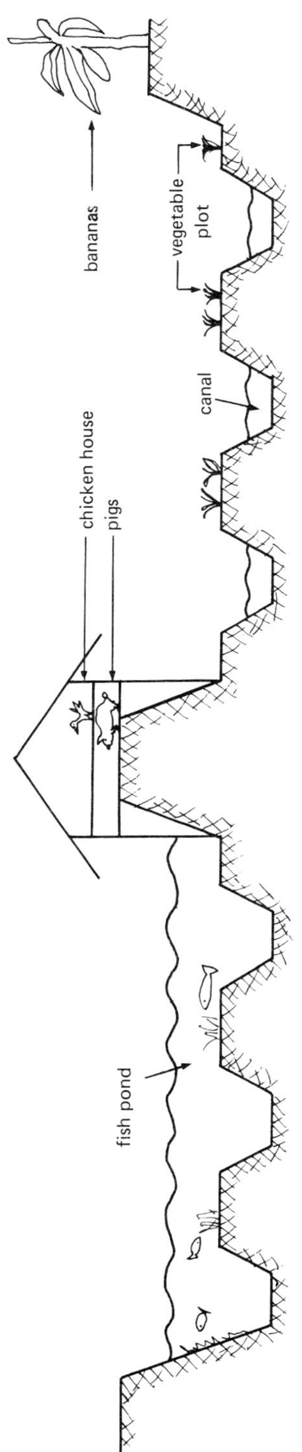

Fig. 12.5. Small-scale integrated farming in Central Thailand.

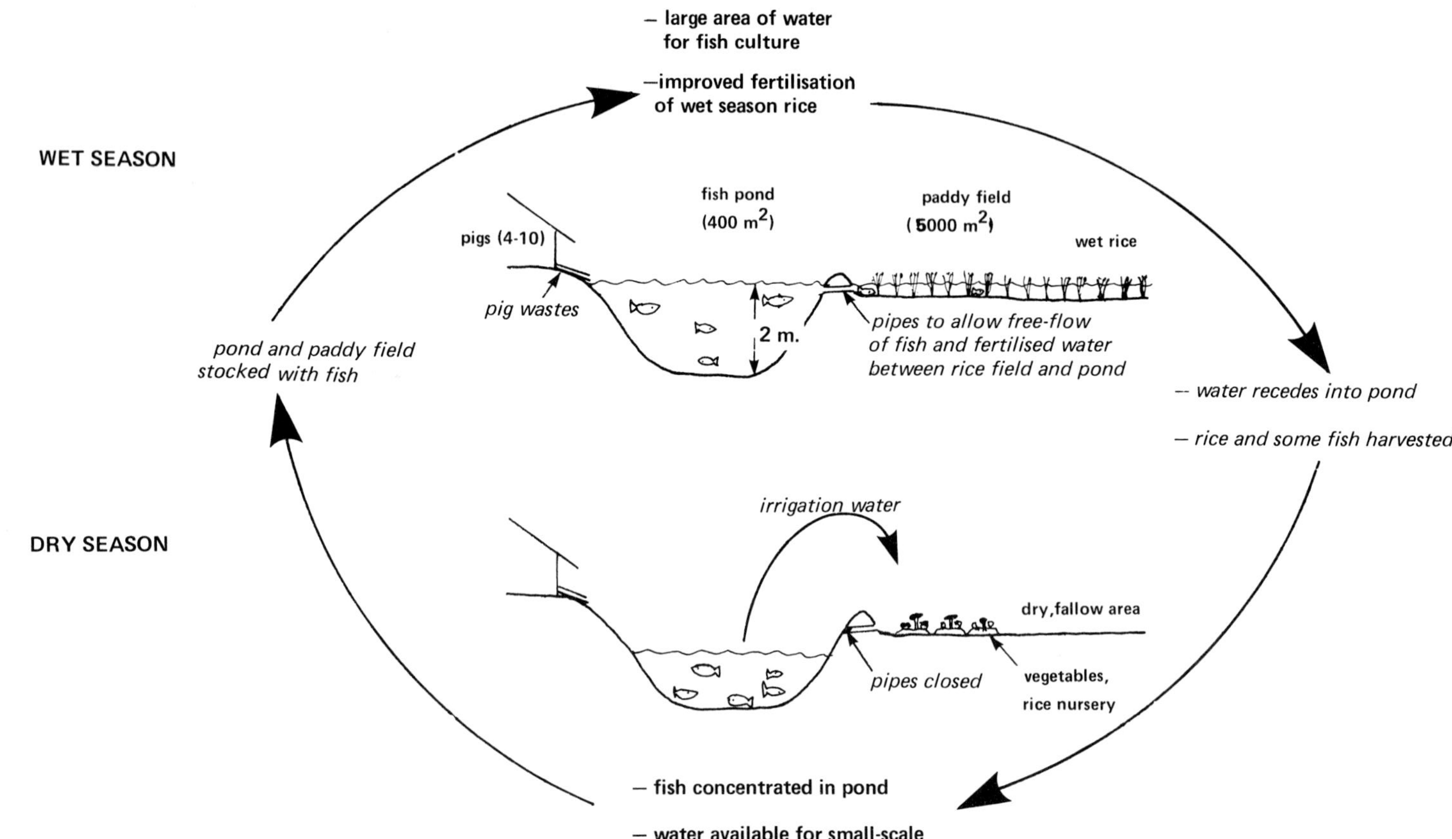

Fig. 12.6. Pond-paddy-pig system in North East Thailand.

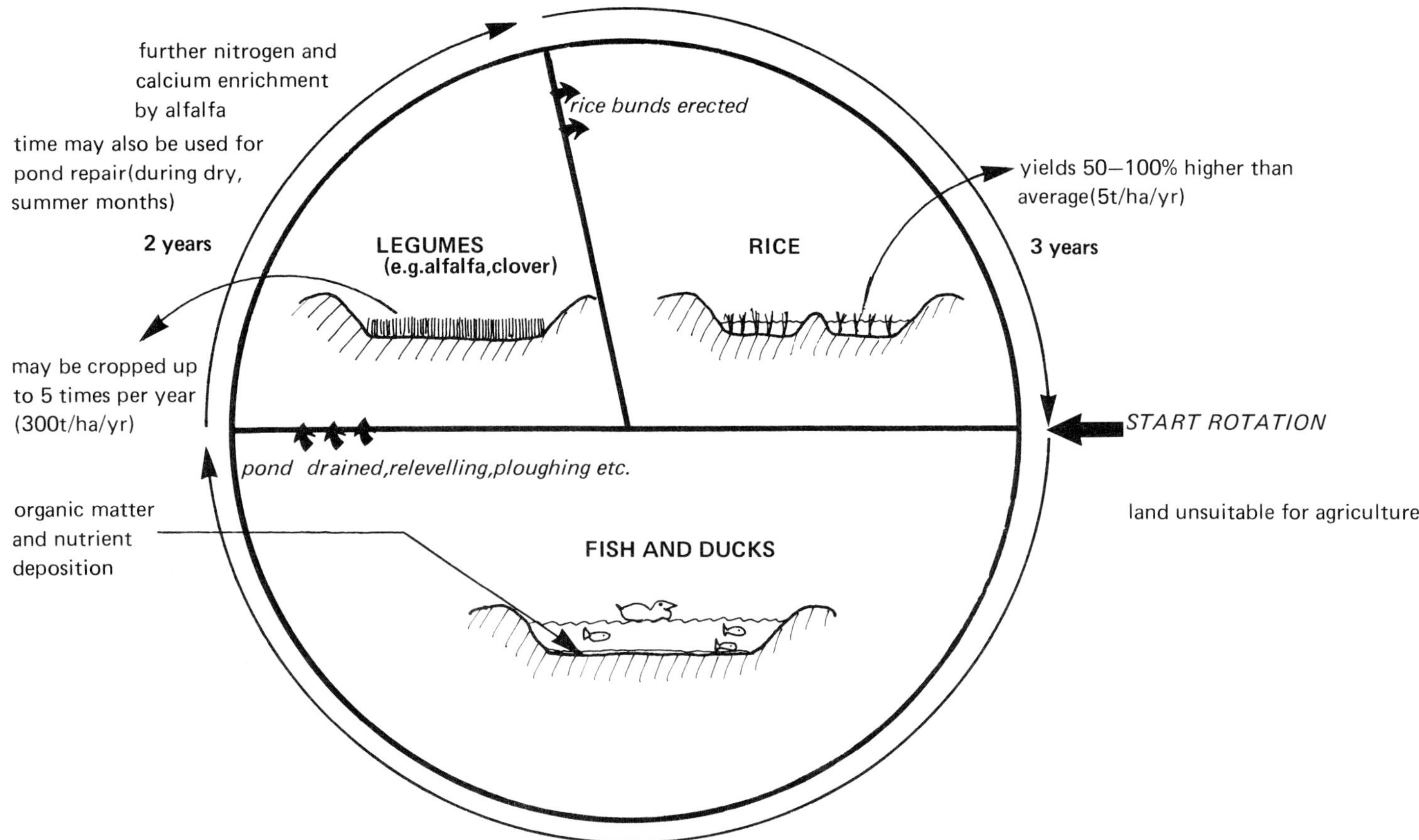

Fig. 12.7. Long-term rotational system in Hungary.

for fish culture and fish pond effluent for the irrigation of maize, revealed the following problems:

— cost of construction and maintenance of high-rate stabilisation pond
— a faster rate of algal production than consumption by fish
— considerable human pathogens remained in fish culture water, and that used for maize irrigation
— a low fertilisation value of the pond water that was used to irrigate the maize.

Thus, Edwards *et al.* (1980) have proposed that it would be simpler and more effective to recycle sewage into fish through a stabilisation pond system composed of earth ponds (Fig. 12.8). Mara (1977) suggested a similar system for the tropics, that could be modified to provide for fish and crop production as well as for waste treatment. The efficiency of algal consumption by fish was however found to increase with detention time of wastes in the system and thus maximum efficiency would require large areas of land to obtain significant fish yields.

12.4 Multi-component systems

Multi-component systems are a further step away from simple integration into more complex and interdependent form of production. Again, these may be seen on a small-scale; perhaps most integrated subsistence systems eventually become more complex as the full utility of a fish pond is realised.

12.4.1 Small-scale (Nonsang, North East Thailand)

The construction of a fish pond in the position illustrated in Fig. 12.9 will have many consequences for the farm's agricultural production, as is shown by the web of relationships that have grown around it. The fish pond for food production is of central importance, but waste disposal, water and nutrient conservation may be just as beneficial where seasonal water shortage occurs. Degradation of upland soils is one unfortunate consequence of the cultivation methods and crops grown, but rice paddy and fish ponds constitute a trap for the rain leached nutrients and those of other human activities. Local processing of primary field crops consumed locally, such as rice, and re-use of by-products for animal and fish systems, is an energy and nutrient efficient strategy.

12.4.2 Large-scale example

(a) Chinese integrated system

The productivity of many Chinese communes and farms is based on the fish pond. The mulberry dyke/carp pond system, reported in the Zhujiang delta in China demonstrated that the fish pond is the 'heart' of the farm (Ruddle *et al.*, 1983); it should be noted that in addition to a reuse and recycling of resources within the farm, large and increasing amounts of feed and stock inputs are

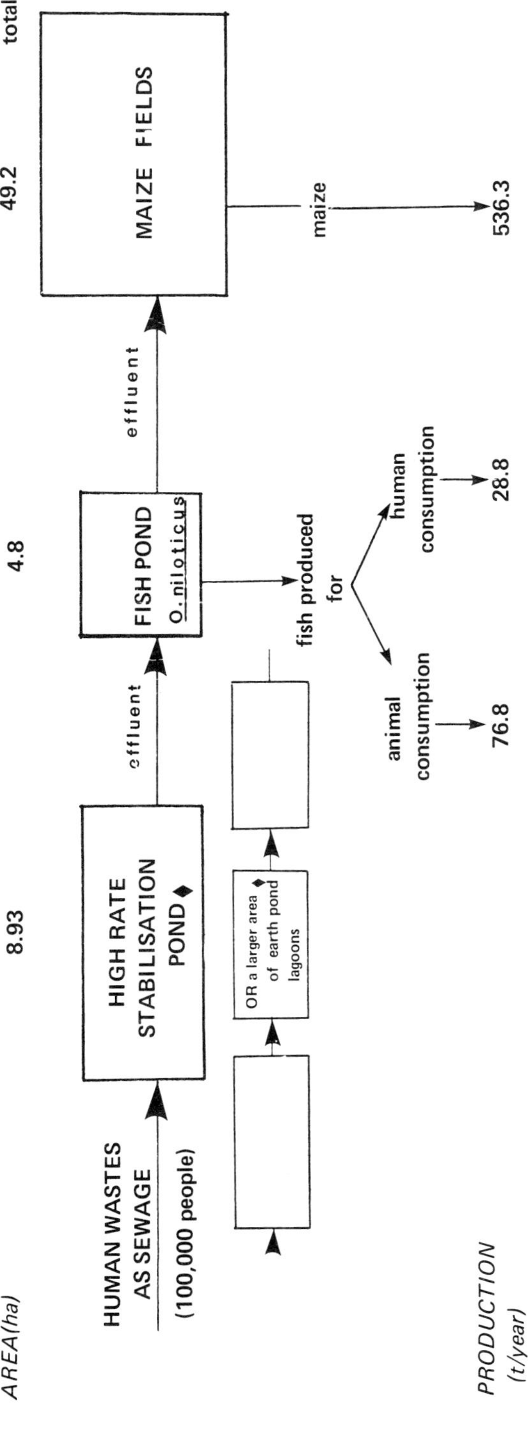

Fig. 12.8. Sewage-fish-maize system using either a high rate stabilisation pond or series of earth lagoons for waste pretreatment.

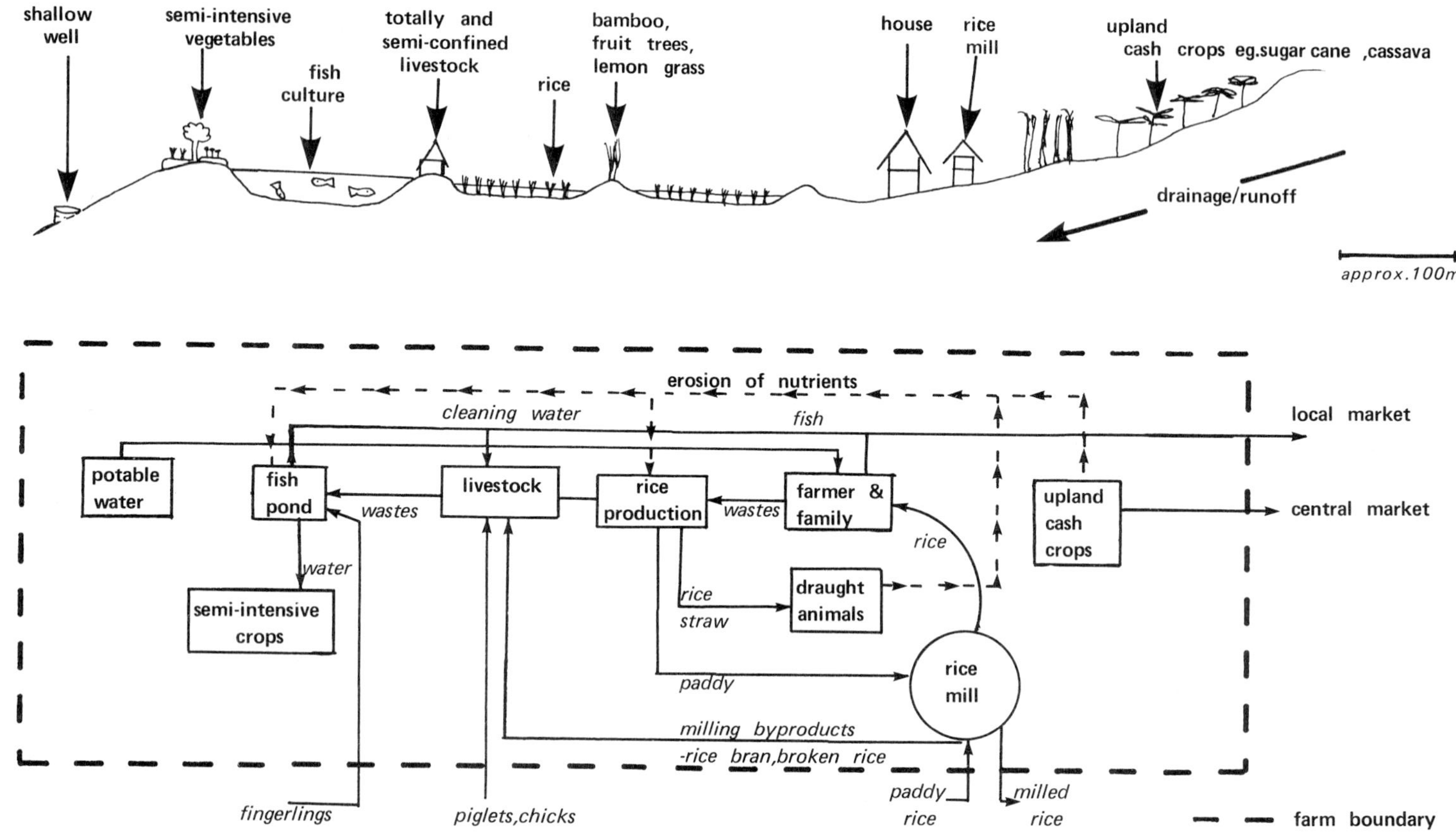

Fig. 12.9. Multi-component system used under rain-fed conditions in North East Thailand.

obtained from outside (Fig. 12.10). Many of these inputs will come from the vicinity of the commune and those that are the most economical to utilise and most readily available will be used. Intrinsic to these systems is a high requirement for labour, and, increasingly, the availability of skilled technicians specia-

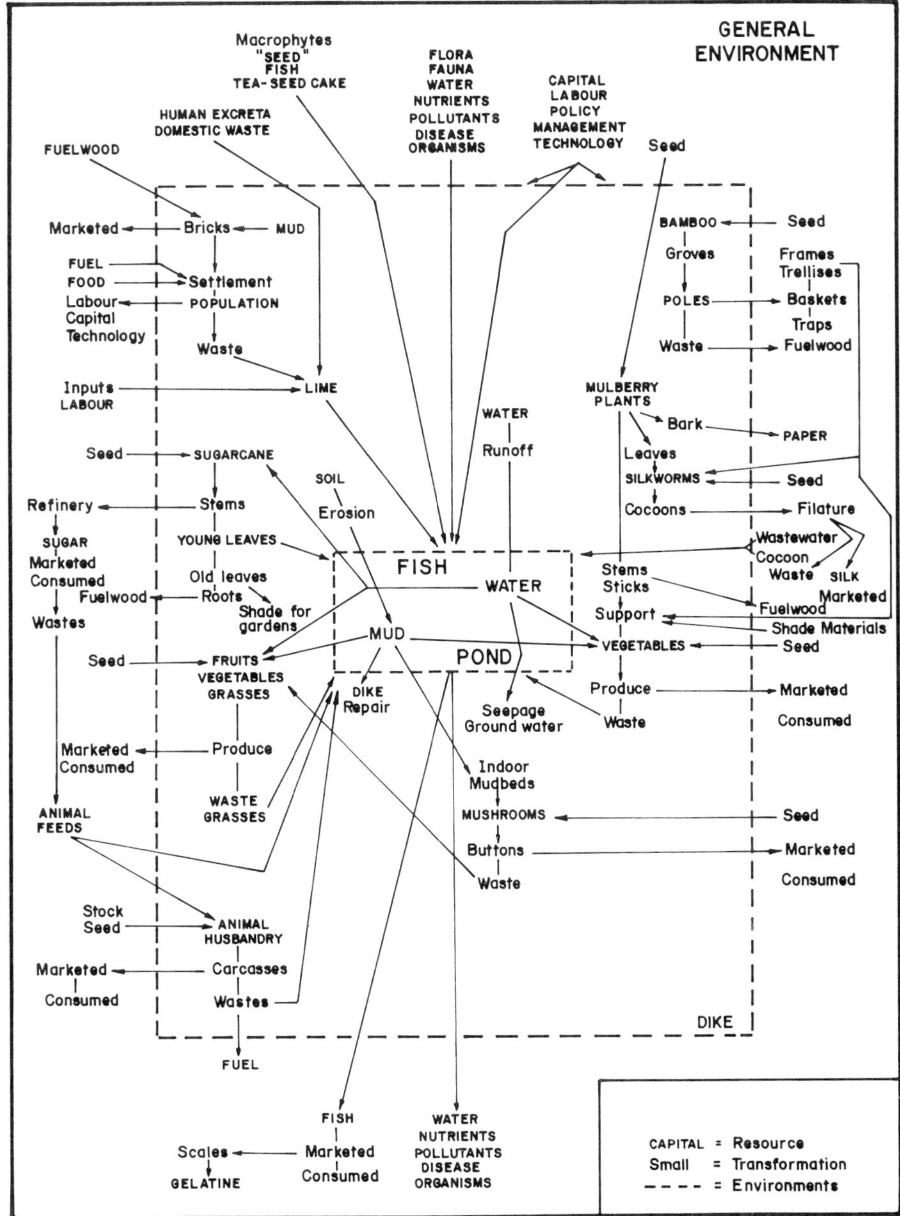

Fig. 12.10. Mulberry dyke/carp pond system in the Zhujiang delta, China. (From Ruddle *et al.*, 1983.)

lised in certain components within the farm (Coche, 1980). Power, in the form of electricity and biogas, is widely available to improve productivity further.

(b) Agro-industrial scale

The highly capitalised large-scale system, as is in operation in Thailand (Fig. 12.11) is likely to become increasingly common. The economic and energy use advantages of onsite feed production, processing and use may be appreciated. In addition, large-scale waste and by-product availability may encourage energy production, to service farm requirements.

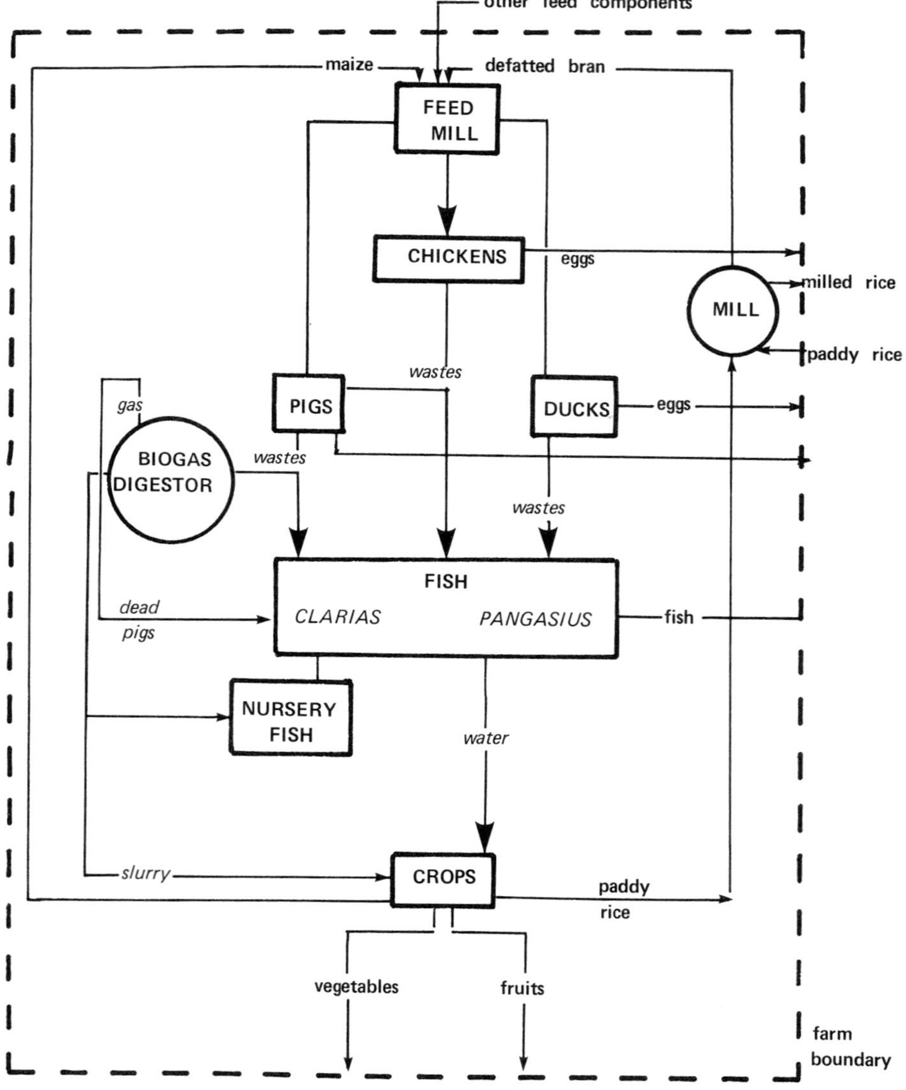

Fig. 12.11. Schematic diagram of Kirikhan Farm, Thailand in which livestock, crops and fish are integrated.

12.5 Potential of integrated systems

12.5.1 Pond and system design

Large-scale, multi-component systems can, with present technology, give positive economic and energetic advantages over intensive systems in many instances. The very limits of pond fish production can be stretched using complete diets and aeration, usually by high applications of energy. These practices are perhaps most at odds with the rationale of integrated farming, not just because inputs are large or expensive in energy terms, but also that recycling of wastes for further use is problematic or uneconomic.

Schroeder (1978) has suggested that higher fish stocking densities and yields than those presently possible using wastes (around 32 kg/fish/ha/day) will require more sophisticated approaches; production of algae and bacteria in isolation from fish feeding might overcome food supply constraints without causing oxygen deficit problems within the culture pond. A system such as this is already used for milkfish (*Chanos chanos*) culture, whereby a 'kitchen' pond is heavily fertilised and its effluent containing large concentrations of natural food is directed into the culture pond. Similarly, the effluent from anaerobic, but highly fertile, catfish ponds in Thailand is sometimes used in larger ponds nearby to produce tilapia and carps.

The other major obstacle to high fish stocking densities and yields in ponds is a build-up of organic and toxic wastes in the sediments. Liao and Chen (1983) have described a Taiwanese system for intensive culture of tilapias which overcomes these problems (Fig. 12.12), by allowing regular solids removal. This system allows the recovery of the fish culture wastes in a concentrated form which can subsequently be re-used as fertilisers in less intensive aquaculture or arable cropping. It has also been shown that a high degree of nitrification occurs to solid and liquid wastes within the culture system, increasing their usefulness further (Rijn *et al.*, 1984).

A further integrated 'intensification' might be feasible if the complete diets presently used could be replaced by a concentrated natural food effluent, especially blue-green algae, from a 'kitchen pond' together with a cheaper supplementary feed (Fig. 12.12). The 'kitchen pond' in turn might be fertilised by the culture pond effluent and/or animal or crop wastes. The concentrated phytoplankton produced in the kitchen pond would efficiently produce oxygen and absorb ammonia. Limiting factors to the system would be the relative size of 'kitchen pond' to culture pond for efficient nutrient removal and the stability of the climate and other local factors which might affect the phytoplankton growth. Water circulation within the 'kitchen pond' would undoubtedly improve the stability and retain high concentrations of microbial and plankton floc in suspension and available as feed for the fish. Further, the 'kitchen pond' might double as a source of highly fertile irrigation water for associated agriculture. This water would be far more concentrated in nutrients than that from a conventional fish pond, where fertility is limited by the oxygen requirements of the fish. Another advantage of this system, especially for arid lands,

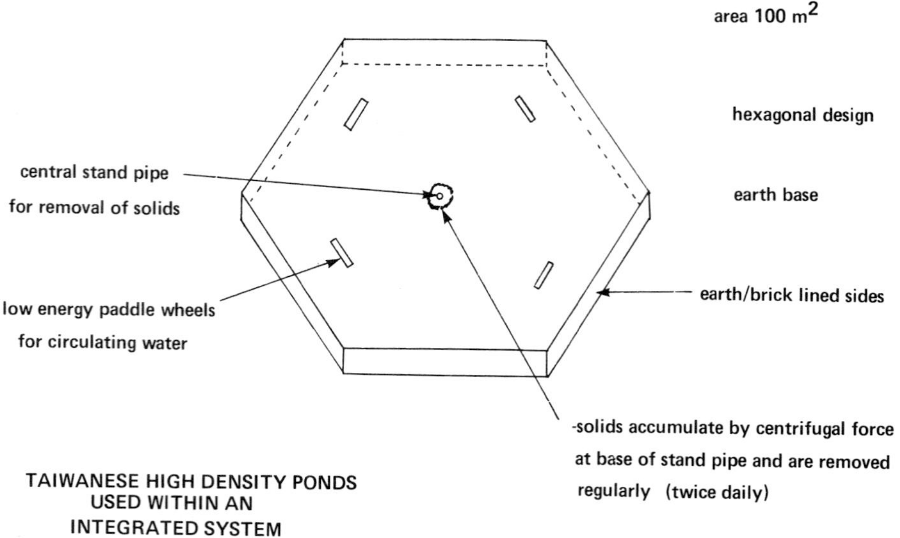

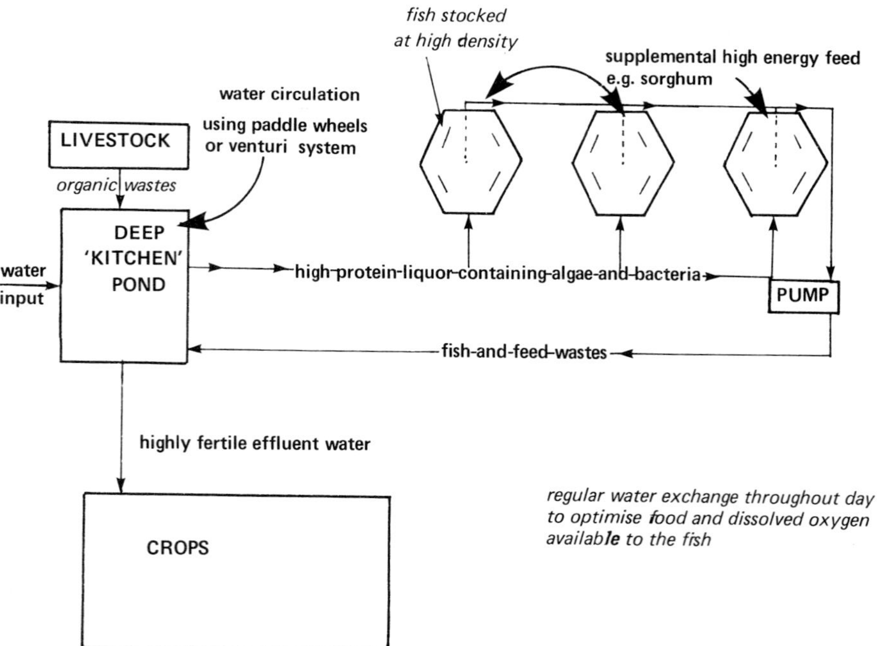

Fig. 12.12. Taiwanese system for intensive tilapia production and its role in an integrated system.

would be the high production achievable in terms of water use. Hepher (personal communication) suggested that only around 4000 m^3 of water is used per tonne of fish produced, compared to around 4400 m^3 in normal pond culture. This compares very favourably with Israeli intensive systems (35000 m^3/t), or other intensive tilapia culture methods, e.g. Baobab Farm; Mombassa

200000 m^3/t). A consequence of this high efficiency of water use would however be the requirement for pumping energy, which might be considerable. Greater advantage of the range of natural foods available and the full pond space, might be taken by the stocking of a bottom feeder. Microphagous tilapia species combined with bottom feeding giant prawns (*Macrobrachium rosenbergii*) are proving to be very productive in pond systems in general. Their well defined and non-conflicting feeding and living requirements would be well suited to the environment of a Taiwanese type pond system as described.

Another potential pathway for future development of integrated systems is the harmonisation of extensive and intensive fish culture as is under development in Hungary. The use of fish process wastes as silage for on-feeding to carnivorous species such as European catfish (*Silurus glanis*), or equivalent warm water species, would considerably reduce economic and energetic costs (Little, 1983). In addition energy costs of semi-intensive fish culture might be reduced or eliminated by energy production on the farm. Biogas is produced on some of the larger integrated farms throughout South East Asia, and the end-use of the gas can be considered for aquaculture or any other of the farm's energy requirements. Its use as a fuel for especially adapted total energy modules, as constructed by several manufacturers (e.g. Fiat, 1979) gives an efficient electrical and thermal output. A system designed to use the thermal output for heating and processing, and the electrical energy for pumping or aeration, etc. is conceivable.

Box 12.1. Sketching systems.

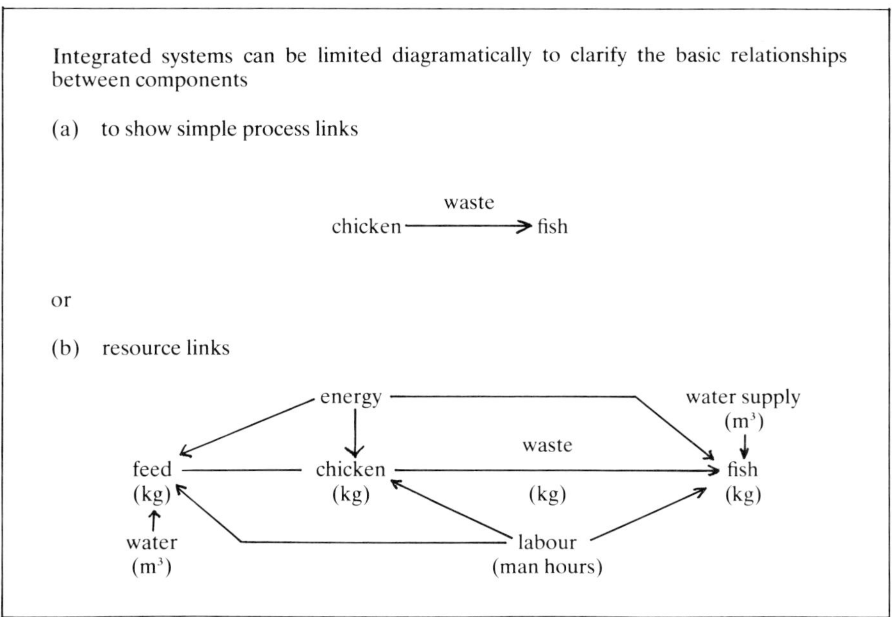

APPENDICES

Appendix 1. Feed formulations for pig farming in the tropics. (Source: Delmendo, 1980.)

	Starter (%)	Growing mash (%)	Fattening mash (%)	Breeding mash (%)	Lactating mash (%)
Corn	35–60	25–60	25–60	15–40	20–50
Corn grits	–	0–15	0–15	0–20	0–15
Corn gluten feed	–	0–15	0–15	0–15	0–15
Pollard	5–15	11–30	0–35	0–20	0–20
Wheat bran	–	5–20	0–15	10–30	10–25
Rice bran (first quality)	–	0–10	0–25	0–15	0–15
Fishmeal	5–10	5–10	–	2–5	5–10
Meat and bone meal	–	0–5	2–7	0–5	0–5
Skim-milk powder	0–20	–	–	–	–
Soybean meal	5–15	2–10	2–5	2–7	5–10
Copra meal	–	0–3	2–7	0–7	0–7
Leucaena glauca leaf meal	0–3	5	5	5	5
Molasses	0–3	5	5	5	5
Sugar	0–10	–	–	–	–
Minerals	1	1·5	1·5	1·5	1·5
Vitamins			as manufacturers' instructions		
Antibiotics			as manufacturers' instructions		
Crude protein	19·0	16·0	14·5	14·5	18·5
Starch equivalent (SE)	70·0	68·0	69·0	64·0	66·0
SE/crude protein	3·7	4·2	4·6	4·4	3·6
Crude fibre (max)	4·0	6·0	6·0	8·5	7·0

Appendix 2. Daily feed quantities for pig farming in the tropics. (Source: Delmendo, 1980.)

Pig weight (kg)	Age (wk)	Daily feed quantity (kg)
15	8	0·75
20	10	0·90
25	12	1·10
30	14	1·30
35	15	1·50
40	17	1·70
45	18	1·80
50	19	2·00
55	20	2·10
60	21	2·30
65	22	2·40
70	23	2·50
75	24	2·60
80	25	2·70
85	26	2·80
90	27	2·90
95	28	3·00
100	29	3·00

Appendix 3. Composition of poultry waste (% dry weight). (After Kerns and Roelofs, 1977.)

Constituent	(%)		Remarks
Crude protein	33·92		Fresh (n = 23)
Moisture	75·00		Fresh (n = 29)
Total nitrogen	6·73		Crude protein = 42·06%
% of total nitrogen	70·2	urinary	61·2% uric acid 9·0% NH_4 salts
	29·8	faecal	Unutilised protein from feed; micro-organisms from gut; feathers
P_2O_5	4·54		
K_2O	2·03		
Ash	20·41		
Organic matter	66·29		

Appendix 4. Production of one day old ducklings.

Production of duckling is dependent on maintaining an adequate broodstock (1000–2000 1–2 yr old ducks); at a sex ratio of 1M:4–6F Peking ducks, 1 ducks lays ~120–160 eggs per season, with 2 eggs/3 days at peak production periods. One duck can produce 70–80 one day old ducklings within an 8 month laying period if continuously supplied with clean drinking water and high protein feed (~16–18% digestible protein, plus calcium carbonate and other minerals), of which 0·42–0·52 kg is required/egg produced.

Each brood duck consumes 40–45 kg feed from first day of hatching up to the start of egg laying, requires 1–1·5 m² of yard area plus 2–4 m² pond area and 0·3–0·5 m² pen area. Egg laying is stimulated by light and can be regulated through artificial light. Two watts of light (preferably yellow) are required per 1 m² of area. (Woynarovich, 1979.)

Appendix 5. Nursing of young ducklings.

Young day-old ducklings require controlled environments (temperature, feed, drinking water, space) up to 10–14 days, after which they can be stocked in the ponds. During the first week of life 50–55 ducklings can be reared per m² within a heated room, with a screen floor (1–5 cm mesh, 2 mm guage) to allow manure and uneaten food to fall through. Pelleted starter feed is provided in demand feeders, with clean tepid water in troughs which are designed to allow access to the beak only, preventing the ducks from getting wet. Air temperature is 30–32°C. After the third or fourth day, ducklings are released into a small enclosed pen during good weather and provided with shallow splashing pools to acclimate them. Special care should be taken to prevent feed sticking to the heads and backs of the ducklings.

Appendix 6. Dietary ingredients used in duck feed formulations. (Source: Woynarovich, 1980.)

Ingredient	Type of feed Starter (%)	Rearing (%)	Layer (%)
Maize	20·0	20·0	20·0
Wheat	57·9	53·9	40·9
Fishmeal (64% crude protein)	10·0	2·0	2·5
Soya grit (47% crude protein)	8·0	9·8	12·3
Meat meal (50% crude protein)	0·9	4·0	5·7
Wheat bran	2·5	8·7	5·0
Food lime	–	0·3	5·3
Methionine premix (5% active ingredient)	–	0·5	0·5
Salt	0·2	0·3	0·2
Vitamin premix	0·5	0·5	0·5
Mineral premix	–	–	2·5
Wheat germ	–	–	1·5
Milled sunflower seed cake (oil extracted)	–	–	3·1

Appendix 7. Some grass species that may be used for pasture and fodder production suitable for goats. (From Devendra and McLeroy, 1982.)

Environment	Common name	Species
Humid tropics	African star or giant star	*Cynodon plectostachyus*
	Bermuda	*Cynodon dactylon*
	Setaria	*Setaria spendida*
	Elephant or Napier	*Pennisetum purpureum*
	Guinea	*Panicum maximum*
	Pangola	*Digitaria decumbens*
	Para	*Brachiaria mutica*
	Signal	*Brachiaria brizantha*
Dry tropics	Buffel	*Cenchrus ciliaris*
	Columbus	*Chloris gayana*
	Makarikari	*Panicum coloratum*
Montane tropics	Bahia	*Paspalum notatum*
	Dallis	*Paspalum dilatatum*
	Kikuyu	*Pennisetum clandestinium*
	Nandi setaria	*Setaria sphacelata*

Appendix 8. Some leaves that are commonly used for feeding goats in the humid tropics. (From Devendra and McLeroy, 1982.)

Region	Common name	Species
Africa	Acacia	*Acacia spp.*
	Amaranthus	*Amaranthus spinosus*
	Baphia	*Baphia rutida*
	Euphorbia	*Euphorbia heterophylla*
	Flemingea	*Flemingea congesta*
	Grifforia	*Grifforia simplicifolia*
	Tridax	*Tridax procumbens*
South East Asia	Anjan	*Handwicbia binnata*
	Babul	*Acacia arabica*
	Banana	*Musa spp.*
	Bargad	*Ficus bengalensis*
	Bari	*Ziziyphus jujuba*
	Cassava	*Manihot esculenta*
	Gliricida	*Gliricida sepium*
	Gular	*Ficus glomerata*
	Hibiscus	*Hibiscus rosa-sinensis*
	Ipil-ipil	*Leucaena leucocephala*
	Jackfruit	*Artocarpus integrifolia*
	Khanthal	*Artocarpus spp.*
	Mulberry	*Morus indica*
	Pigeon pea	*Cajanus cajan*
	Pipal	*Ficus religiosa*
	Neem	*Azadirachta indica*
	Siras	*Albizia lebbeck*

Appendix 9. Legume plant residues for use as fish feeds. (After FAO, 1983.)

Common name	Scientific name	Part of plant used	Crude protein (N × 6·25)	Lipid	Carbohydrate	Notes
			as % DM			
Chick pea	*Cicer arietinum*	Fresh aerial parts	11·3	2·2	47·9	Found in India and other parts of the tropics
Lupins	*Lupinus* sp.	Fresh aerial parts	26·6	2·6	37·8	Found in cooler tropics
Saman or cow tamarind	*Samanea saman*	Fresh leaves	22·1	7·0	35·5	High in tannins
		Seeds	20·7	3·6	55·5	Fed experimentally to native fish in Venezuela (*Cachama* sp.) at moderate levels
Mung bean	*Phaseolus mungo*	Fresh aerial parts	19·4	2·5	35·3	
		Seeds	24·4	1·0	65·8	Absence of glucosides–believed safe at moderate levels
Horse grain	*Duliochos bithomus*	Fresh aerial parts	17·6	2·2	51·7	Herb in tropical Asia
Cow pea	***Vigna nuguiculata***	Fresh aerial parts	30·6	1·8	28·9	

REFERENCES

Afinowi, M. A. and Ezenwa, B. I. (1982). The utilisation of coastal areas for aquaculture development in Nigeria. *CIFA Tech. Pap., 9,* 67–80.

Alabaster, J. S. and Scott, B. (1967). Grass carp (*Ctenopharyngodon idella* Val.) for aquatic weed control. *Proc. European Weed Res. Counc. Int. Symp. Aquatic Weeds, 2,* 123–126.

Allen, G. H. and Hepher, B. (1979). Recycling of water through aquaculture and constraints to wider application. In: T. V. R. Pillay and W. A. Dill (eds), *Advances in Aquaculture,* Fishing News Books, Farnham, England, pp. 478–487.

Almazan, G. and Boyd, C. E. (1978). Effects of nitrogen levels on rates of oxygen consumption during decay of aquatic plants. *Aquat. Bot., 5,* 119–126.

Anon (1983). Sheep and goats in developing countries. A World Bank Technical Paper *ISSN 0253–7494.* Winrock International Livestock Research and Training Center II Series.

Anon (1984). Fish and lotus. *Fish Farming Int.*

Anon (1985[a]). The golden wastes of Kirikarn. *Pig International.* pp. 12–14.

Anon (1985[b]). Shrimp culture in the semi-saline zone of the delta. *Tech. Report 13, 2, Tech. Report of the Delta Development Project Bangladesh-Netherlands Joint Program under BWDB,* Arnhem, The Hague. 102 pp.

Ardiwinata, R. O. (1957). Fish culture on paddy fields in Indonesia. *Proc. Indo-Pacific Fish. Counc., 7 (*II *and* III*),* 119–162.

Ark (1959). Manuring of fish ponds. *Agric. Pak., 10,* 122–135.

Armbrester, W. (1972). The growth of caged *Tilapia aurea* (Steind.) in fertile farm ponds. *Proc. Annu. Conf. Southeast Assoc. Game Fish Comm., 25,* 446–451.

Avault, J. W., Smitherman, R. O. and Shell, E. W. (1968). Evaluation of eight species of fish for aquatic weed control. *FAO Fish. Rep., 44 (5),* 109–122.

Bagnall, L. O., Furmantdes, S., Hentges, J. R., Nolan, N. J. and Shirley, R. L. (1974). Feed and fiber from effluent grown water hyacinth. In: *Wastewater Use in the Production of Food and Fiber.* Proceedings EPA–660/2–74–041. Washington, D.C.

Bain, R. (1982). An appraisal of current and potential replacement of agricultural feeds for rainbow trout culture. Ph.D. thesis. University of Stirling, Scotland.

Barasch, H. and Schroeder, G. L. (1984). Use of fermented cow manure as a feed substrate for fish polyculture in stagnant water ponds. *Aquaculture, 36,* 127–140.

Barasch, H., Plavnik, I. and Moav, R. (1982). Integration of duck and fish farming: experimental results. *Aquaculture, 27,* 129–140.

Barnett, A., Pyle, L. and Subramanian, S. K. (1978). *Biogas Technology in the Third World: A Multidisciplinary Review.* International Development Research Centre, Ottawa, Canada. 132 pp.

Bayne, D. R., Dunseth, D. and Ramirios, G. G. (1976). Supplemental feeds containing coffee pulp for rearing tilapia in Central America. *Aquaculture, 7,* 133–146.

Beveridge, M. C. M. (1984). Cage and pen fish farming. Carrying capacity models and environmental impact. *FAO Fish. Tech. Pap., 255,* 131 pp.

Beveridge, M.C.M. and Muir, J. F. (1987). Cage culture in S. E. Asia—a review. Proceedings of International Conference on Limnology. Kuala Lumpur, Malaysia, December 1982. *Arch. Hydrobiol.* (in press).

Bitterlich, G. (1985). The nutrition of stomachless phytoplanktivorous fish in comparison with tilapia. *Hydrobiologia, 121,* 173–179.

Bitterlich, G. and Graiger, E. (1984). Phytoplanktivorous or omnivorous fish? Digestibility of zooplankton by silver carp. *Aquaculture, 40,* 261–263.

Bondari, K. and Shepherd, D. C. (1981). Soldier fly larvae as feed in commercial fish production. *Aquaculture, 24,* 103–109.

BOWEN, S. H. (1976). Mechanism for digestion of detrital bacteria by the cichlid fish *Saratherodon mossambicus* (Peters). *Nature (Lond.), 260,* 137–138.

BOWEN, S. H. (1982). Feeding, digestion and growth—qualitative considerations. In: R. S. V. Pullin and R. H. Lowe McConnell (eds), *The Biology and Culture of Tilapias. ICLARM Conf. Proc., 7,* 141–156.

BOYD, C. E. (1968). Freshwater plants: A potential source of protein. *Econ. Bot., 22,* 359–368.

BOYD, C. E. (1976). Accumulation of dry matter, nitrogen and phosphorus by cultivated water hyacinths. *Econ. Bot., 30,* 51–56.

BROUGH, O. L. (1974). Livestock production in the Near East region—an overview. Regional Workshop for Sheeps and Forage Production, Feb. 1974. Beirut, Lebanon.

BRYAN, F. L. (1977). Diseases transmitted by foods contaminated by waste water. *J. Food Prot., 40,* 45–56.

BUCK, D. H., BAUR, R. J. and ROSE, C. R. (1979). Experiments in recycling swine manure in fish ponds. In: T. V. R. Pillay and W. A. Dill (eds), *Advances in Aquaculture,* Fishing News Books, Farnham, England.

BURAS, N. L. and NIV DUEK, S. (1985). Reactions of fish to micro-organisms in wastewater. *Appl. and Environ. Microbiol., 50 (4),* 989–995.

BURNS, R. P. and STICKNEY, R. R. (1980). Growth of *Tilapia aurea* in ponds receiving poultry wastes. *Aquaculture, 20,* 117–121.

BUSCH, C. D. and GOODMAN, R. K. (1981). Water circulation—an alternative to emergency aeration. *J. World Maricul. Soc., 12 (1),* 13–19.

BUVANENDRAN, V. and SIRIWARDENA(1970). Rubber seed meal in poultry diets. *Ceylon Vet. J., 81,* 33–38.

CARVALHO, J. N. de and FERNANDES, J. A. (1979). Intensive raising of fish in two earthen ponds of the DNOCS. *Bol. Tech. Dep. Nac. Obras Contra Secas, 36 (1),* 15–20.

CHALMERS, M. I. (1974). Studies on the rumen microbial activity of buffalo and Zebu cattle. Concentrations of micro-organisms and total and particulate nitrogen in the rumen liquor. *Indian J. Anim. Sci., 40,* 600–609.

CHAO, CHIA-HSING (1983). Cultivation of the duckweed *Spirodella* on cesspool slurry for fish feed. M.Sc. thesis, Asian Institute of Technology, Thailand.

CHEN, T. P. (1954). The culture of tilapia in rice paddies in Taiwan. Jt. Comm. Rural Reconstr. *China Fish Ser., 2,* 29 pp.

CHEN, T. P. and LI, T. (1980). Integrated agriculture/aquaculture studies in Taiwan. In: R. S. V. Pullin and Z. H. Shehadeh (eds), *Integrated Agriculture-Aquaculture Farming Systems. ICLARM Conf. Proc., 4,* 239–241.

CHRISTENSEN, M. S. (1981). Preliminary tests on the suitability of coffee pulp in the diets of common carp (*Cyprinus carpio*) and catfish (*Clarias mossambicus*). *Aquaculture, 25,* 235–242.

CLARKE, R. (1983). Methods for the removal of the suspended solid fraction from fish farm effluent with particular reference to the swirl concentrator. M.Sc. thesis, University of Stirling, Scotland.

CLOETE, T. E., TOERIEN, D. F. and PIETERSE, J. H. (1984). The bacteriological quality of water and fish of a pond system for the treatment of cattle feedlot effluent. *Agricultural Wastes, 9,* 1–15.

COCHE, A. G. (1980). Freshwater aquaculture development in China—rep. FAO/UNDP study tour organised for French-speaking African countries 22/4–20/5/80. *FAO Fish. Tech. Pap., 215.*

COCHE, A. G. (1982). Cage culture of tilapias. In: R. S. V. Pullin and R. H. Lowe-McConnell (eds), *The Biology and Culture of Tilapias. ICLARM Conf. Proc., 7,* 205–245.

COCHE, A. G. (1983). The cultivation of fish in cages. An indexed bibliography 1965–1983. *FAO Fish. Circ., 714, Rev. 1,* 61 pp.

COCKRILL, W. R. (1974). *The Husbandry and Health of the Domestic Buffalo.* FAO, Rome. 993 pp.

COE, W. B. and TURK, M. *Processing Animal Waste by Anaerobic Fermentation.* Hamilton Standard, Division of United Aircrafts Corp. USA.

COHEN, D. and RA'ANAN, Z. (1983). The production of the freshwater prawn, *Macrobrachium rosenbergii* in Israel. III Density effect of all-male tilapia hybrids on prawn yield characters in polyculture. *Aquaculture, 35,* 57–71.

COLMAN, J. A. and EDWARDS, P. (1985). Feeding pathways and environmental constraints in waste-fed aquaculture: balance and optimisation. Paper presented at the Bellagio Symposium, Italy on Detritus and Aquaculture, August 1985. 42 pp.

COOK, C. D. K., GUT, B. J., RIX, E. M., SCHNELLER, J. and SEITZ, M. (1974). *Water Plants of the*

World. A Manual for the Identification of the Genera of Freshwater Macrophytes. D. & W. Junk B.V., The Hague, Netherlands.

Creswell, D. C. and Kompiang, I. P. (1981). Studies on snail (*Achatina fulica*) meal as a protein source for chickens: 1. Chemical composition, metabolisable energy and feeding value for broilers. *Poult. Sci., 60* (8), 1854–60.

Crocker, S. A. (1983). To study ways of adding value to animal wastes. Report on a visit to China and USA to the Nuffield Farming Scholarships Trust. 90 pp.

Cross, D. G. (1969). Aquatic weed control using grass carp. *J. Fish Biol., 1,* 27–30.

Cruz, E. M. and Laudencia, I. L. (1978). Screening of feedstuffs as ingredients in the rations of Nile tilapia. *Kalikasan, 7 (2),* 159–164.

Cruz, E. M. and Shehadeh, Z. H. (1980). Preliminary results of integrated pig-fish and duck-fish production tests. In: R. S. V. Pullin and Z. H. Shehadeh (eds), *Integrated Agriculture-Aquaculture Farming Systems. ICLARM Conf. Proc., 4,* 225–238.

Dale, J. T. (1979). World bank shifts focus on Thailand World Sanitation projects. *J. Water Pollut. Control Fed., 51,* 662–665.

De Bout, A. (1955). Premier essai de rhizipisciculture etc. *Bull Agric. Congo Belge, LXVII,* 6.

Degani, G., Dosoretz, C., Levanon, D., Marchaim, U. and Perach, Z. (1982). Feeding *Saratherodon aureus* with fermented cow manure. *Bamidgeh, 34,* 4.

De La Cruz, C. R. (1980). Integrated agriculture-aquaculture systems in the Philippines, with two cases on simultaneous and rotational rice-fish culture. In: R. S. V. Pullin and Z. H. Shehadeh (eds), *Integrated Agriculture-Aquaculture Farming Systems. ICLARM Conf. Proc., 4,* 209–223.

De La Hunt, T. E. (1981). Intensive beef production in Zimbabwe, based on sugarcane by-products. In: A. J. Smith and R. G. Gunn (eds), *Intensive Animal Production in Developing Countries. Occ. Pub. No. 4 British Soc. of Animal Production,* 355–362.

Delmendo, M. N. (1980). A review of integrated livestock-fowl-fish farming systems. In: R. S. V. Pullin and Z. H. Shehadeh (eds), *Integrated Agriculture-Aquaculture Farming Systems. ICLARM Conf. Proc., 4,* 59–71.

Delmendo, M. N. and Gedney, R. H. (1974). Laguna de Bay fish pen aquaculture development. Philippines, *Proc. Annu. Meet. World Maricult. Soc., 7,* 257–265.

De Silva, S. S. and Perera, M. K. (1983). Digestibility of an aquatic macrophyte by the cichlid *Etroplus suratensis* (Bloch) with observations on the relative merits of three indigineous components as markers and daily changes in protein digestibility. *J. Fish. Biol., 23,* 675–684.

De Silva, S. S., Maitipe, P. and Cumaranatunge, R. T. (1984). Aspects of the biology of the euryhaline Asian cichlid, *Etroplus suratensis. Environ. Biol. Fishes, 10 (1/2),* 77–87.

Devendra, C. and Fuller, M. F. (1979). *Pig Production in the Tropics.* Oxford University Press, UK.

Devendra, C. and McLeroy, G. B. (1982). *Goat and Sheep Production in the Tropics.* Longman, UK.

Disney, J. G. and James, D. (1980). Fish silage production and its use. Papers presented at the Indo-Pacific Fisheries Commission Workshop on Fish Silage Production and Its Use. Jakarta, Indonesia, September 17-21, 1979. *FAO Fish. Rep., 230.*

Djajadiredja, R., Jangkaru, Z. and Junus, M. (1980). Freshwater aquaculture in Indonesia with special reference to small-scale agriculture-aquaculture farming systems in West Java. In: R. S. V. Pullin and Z. H. Shehadeh (eds), *Integrated Agriculture-Aquaculture Farming Systems. ICLARM Conf. Proc., 4,* 143–165.

Edie, H. and Ho, B. C. (1979). *Ipomea aquatica* as a vegetable crop in Hong Kong. *Econ. Bot., 23,* 32–36.

Edwards, P. (1980[a]). The production of microalgae on human wastes and their harvest by herbivorous fish. In: G. Shelef and G. J. Soeder (eds), *Algae Biomass.* Elsevier/North Holland Biomedical Press, Netherlands.

Edwards, P. (1980[b]). A review of recycling organic wastes into fish with emphasis on the tropics. *Aquaculture, 21,* 261–279.

Edwards, P. (1980[c]). Food potential of aquatic macrophytes. *ICLARM Stud. Rev., 5,* 51 pp.

Edwards, P. (1982[a]). Report of consultancy at the Regional Lead Centre in China for Integrated Fish Farming. NACA Head Office, NIFI, Kasetsart University, Bangkhen, Bangkok, Thailand. 104 pp.

Edwards, P. (1982[b]). Integrated fish farming in Thailand. *ICLARM Newsl., 5,* 3.

Edwards, P. (1983). The future potential of integrated farming systems in Asia. Proc. V WCAP, 1, 273–281.

EDWARDS, P. (1984). A scheme to recycle septage into high protein animal feed. Paper presented at International Seminar on Resource Recovery and Utilisation (Liquid and Solid Waste, Septage), Shanghai, PRC. November 3–10, 1984.

EDWARDS, P. (1985[a]). Pigs over fish ponds. *Pig International,* pp. 8–10.

EDWARDS, P. (1985[b]). Duck/fish integrated farming systems. Paper presented at 'Duck Production Science and World Practice', November 1985, Cipanas, Bogor, Indonesia.

EDWARDS, P. (1985[c]). Use of terrestrial vegetation and aquatic macrophytes in aquaculture. Paper presented at the Bellagio Symposium on Detritus and Aquaculture, August 1985. 22 pp.

EDWARDS, P. (1985[d]). Aquaculture: a component of low cost sanitation technology. *World Bank Tech. Pap., 36,* 45 pp.

EDWARDS, P. and KAEWPAITOON, K. (1982). Integrated fish farming in Thailand. *ICLARM Newsl., 5,* 3–4.

EDWARDS, P. and SINCHUMPASAK, O. (1981). The harvest of microalgae from the effluent of a sewage fed high rate stabilisation pond by *Tilapia nilotica.* Part 1: description of the system and the study of the high rate pond. *Aquaculture, 23,* 83–105.

EDWARDS, P., SINCHUMPASAK, O., LABHSETWAR, V. K., OUANO, E. A. R. and TABUCANON, M. (1980). Fish cultivation in sewage stabilisation pond effluent. *AIT Research Report, 110,* AIT, Bangkok, Thailand. 206 pp.

EDWARDS, P., SINCHUMPASAK, O. and TABUCANON, M. (1981[a]). The harvest of microalgae from the effluent of a sewage fed high rate stabilisation pond by *Tilapia nilotica.* Part 2: studies of the fish ponds. *Aquaculture, 23,* 107–149.

EDWARDS, P., SINCHUMPASAK, O., LABHSETWAR, V. K. and TABUCANON, M. (1981[b]). The harvest of microalgae from the effluent of a sewage fed high rate stabilisation pond by *Tilapia nilotica.* Part 3: maize cultivation experiment, bacteriogical studies and economic assessment. *Aquaculture, 23,* 149–170.

EDWARDS, P., POLPRASERT, C., PACHARAPRAKITI, C., RAJPUT, V. S. and SUTHIRAWUT, S. (1983[a]). Compost as fish feed, a practical application of detritivory for the cultivation of tilapia. *Aquaculture, 32,* 409–413.

EDWARDS, P., WEBER, K., MCOY, E., CHANTACHAENG, C., PACHARAPRAKITI, C., KAEWPAITOON, K. and NITSMER, S. (1983[b]). Small-scale fishery project in Pathumthani Province, Central Thailand: A socio-economic and technological assessment of status and potential. *AIT Research Report, 158,* AIT, Bangkok, Thailand. 256 pp.

EDWARDS, P., PACHARAPRAKITI, C., KAEWPAITOON, K., RAJPUT, V. S., RUAMTHAVEESUB, P., SUTHIRAWUT, S., YOMJINDA, M. and CHAO, C. H. (1984). Reuse of cesspool slurry and cellulose agricultural residues for fish culture. *AIT Research Report, 166,* AIT, Bangkok, Thailand. 338 pp.

EDWARDS, P., KAMAL, M. and WEE, K. L. (1985). Incorporation of composted and dried water hyacinth in pelleted feed for tilapia, *Oreochromis niloticus* (Peters). *Aquacult. and Fish. Manage., 1,* 233–248.

EDWARDS, P., KAEWPAITOON, K, MCCOY, E. W. and CHANTACHAENG, C. (1986). Pilot small-scale crop/livestock/fish integrated farm. *AIT Research Report, 184,* AIT, Bangkok, Thailand. 131 pp.

EDWARDSON, W. (1976). Fish farming. Report 6: Systems Analysis Research Unit, Energy Studies Unit. University of Strathclyde, Scotland.

ELEMELE, H. O., LAO, D. R. and CHAWAN, C. B. (1980). Evaluation of rabbit excreta as an ingredient in broiler diets. *Br. Poult. Sci., 21,* 345–349.

ESTORES, R. A., LAIGO, F. M. and ADORDIONISIO, C. I. (1980). Carbofuran in rice-fish culture. In: R. S. V. Pullin and Z. H. Shehadeh (eds), *Integrated Agriculture-Aquaculture Farming Systems. ICLARM Conf. Proc., 4,* 53–57.

EUSEBIO, J. A., RABINO, B. I. and EUSEBIO, E. C. (1978). Recycling system in integrated plant and animal farming. *University of the Philippines Tech. Bulletin, 1,* 30 pp.

FALKENMARK, M. (1984). New ecological approach to the water cycle: ticket to the future. *Ambio, XIII (3),* 152–160.

FAO (1977[a]). China: recycling of organic wastes in agriculture. *FAO Soils Bulletin, 40,* 107 pp.

FAO (1977[b]). Freshwater fisheries and aquaculture in China. *FAO Fish. Tech. Pap., 168,* 84 pp.

FAO (1980). Freshwater aquaculture development in China. Report of the FAO/UNDP study tour organised for French-speaking African countries. *FAO Fish. Tech. Pap., 215,* 124 pp.

FAO (1983). Fish feeds and feeding in developing countries—an interim report on the ADCP Feed Development Programme. *ADCP/REP/83/18,* 97 pp.

FEACHAM, R., McGARRY, M. G. and MARA, D. (eds), *Water, Wastes and Health in Hot Climates.* John Wiley and Sons, N.Y., USA.

FIAT (1979). *Totem—Totel Energy Module.* Fiat Auto Publication, Turin, Italy.

GALAPITAGE, D. C. (1982). Economics of cage culture of tilapia in Sri Lanka. In: Aquaculture Economics Research in Asia. Proceedings of a Workshop held in Singapore, June 2-5, 1981, IDRC, Ottawa, Ontario, 128 pp.

GARCIA, R. P. (1979). Cage culture: a promising alternative. *Fish. Today (Philippines), 2* (3), 32–35.

GHOSH, S. K., MANDAL, B. K. and BORTHAKUR, D. N. (1984). Effects of feeding rates on production of common carp and water quality in paddy-cum-fish culture. *Aquaculture, 40,* 97–101.

GODDARD, I. G. H. (1981). Intensive poultry production in Jordan. In: A. J. Smith and R. G. Gunn (eds). *Intensive Animal Production in Developing Countries. Ocean. Pub. Br. Soc. Anim. Prod., 4,* 291–298.

GOHL, B. (1980). *Tropical Feeds.* (2nd edition). FAO Feeds Information Centre, Rome, Italy.

GORBACH, S., HAARRING, R., KNAUF, W. and WERNER, H. J. (1971). Residue analysis in the water system of East Java (River Brantas, ponds, sea water) after continued large scale application of Thiodan in rice. *Bull. Environ. Contam. Toxicol., 6* (11), 40–47.

GOTAAS, (1956). *Composting: Sanitary Disposal and Reclamation of Organic Wastes.* WHO, Geneva, Switzerland.

GRANDSTAFF, T. B. and GRANDSTAFF, S. W. (1986). Choice of rice technology: a farmer perspective. In: D. C. Korten (ed), *Community Management, Asian Experience and Perspectives.* Kumarian Press.

GUERRERO, R. D. (1980). Studies on the feeding of *Tilapia nilotica* in cages. *Aquaculture, 20,* 169–175.

GUERRERO, R. D. (1981). The culture and use of *Perionyx excavatus* as a protein resource in the Philippines. Darwin Centenary Symposium on Earthworm Ecology. Institute Terrestrial Ecology. UK August 30-September 4.

HEIDINGER, R. C. (1971). Use of ultraviolet light to increase the availability of aerial insects to caged bluegill sunfish. *Prog. Fish-Cult., 33,* 187–192.

HEPHER, B. (1952). Rotation of crops in fish ponds. *Bamidgeh, 4* (1/2), 18–19.

HEPHER, B. (1962). Primary production in fish ponds and its application to fertilisation experiments. *Limnol Oceanogr., 7,* 131–136.

HEPHER, B. (1975). Supplementary feeding in fish culture. In: *Proc. 9th International Congress on Nutrition, Mexico* 1972 *Vol.* 3. S. Karger Publ. New York. pp. 183–198.

HEPHER, B. and PRUGININ, Y. (1981). *Commercial Fish Farming, with Special Reference to Fish Culture in Israel.* John Wiley and Sons, N.Y., USA.

HEPHER, B. and SCHROEDER, G. L. (1977). Wastewater utilisation in Israeli aquaculture. In: E. D'Itri (ed), *Wastewater Renovation and Reuse.* Marcel Dekker, New York and Basel. pp. 529–56.

HEPHER, B., SANDBANK, E. and SHELEF, G. (1979). Alternative protein sources for warmwater fish diets. In: J. E. Halver and K. Tiews (eds), *Finfish Nutrition and Fishfeed Technology.* Heinemann, UK. pp. 327–341.

HICKLING, C. F. (1960). Observations on the growth rate of the Chinese grass carp. *Malays. Agric. J., 43,* 49–53.

HICKLING, C. F. (1962). *Fish Culture.* Faber and Faber, London.

HILLIARD, BEARD and PEARCE (1979). Utilisation of piggery waste. (1) The chemical composition and in vitro organic matter digestibility of pig faces from commercial piggeries in south eastern Australia. *Agric. Environ., 4,* 171–180.

HILTON, J. W. (1983). Potential of freeze-dried worm meal as a replacement for fish meal in trout diet formulations. *Aquaculture, 32,* 277–283.

HOLM, L. G., WELDON, L. W. and BLACKBURN, R. D. (1969). Aquatic weeds. *Science, 66,* 699–709.

HOPKINS, K. D. (1983). Tilapia culture in arid lands. *ICLARM Newsl., 6,* 8–9.

HOPKINS, K. D. and CRUZ, E. M. (1982). The ICLARM-CLSU integrated animal-fish farming project: final report. *ICLARM Tech. Rep., 5,* 96 pp.

HOPKINS, K. D., CRUZ, E. M., HOPKINS, M. L. and CHONG, K. C. (1980). Optimum manure loading rates in tropical freshwater fish ponds receiving untreated piggery wastes. *ICLARM Tech. Rep., 2,* 15–29.

HORA, S. L. and PILLAY, T. V. R. (1962). Handbook of fish culture in the Indo-Pacific region. *FAO Fish. Biol. Tech. Pap., 14,* 204 pp.

HUBBERT, R. M. (1983). Effect of anaerobically digested animal manure on the bacterial flora of common carp (*C. carpio*). *Bamidgeh, 35,* 73–79.

HUISMAN, E. A. (1983). Problem into protein—The Egyptian experience. International Agricultural Development, May/June 20-22.

HUISMAN, E. A. (1984). Recent advances in aquaculture and their implications in the African region. In: H. Rosenthal and E. Grimaldi (eds), *Proceedings International Fish Farming Conference, Acquacoltura, 84,* Verona, Italy. (in press).

HUTAGALUNG, R. I. (1981). The use of the crops and their by-products for intensive animal production. In: A. J. Smith and R. G. Gunn (eds), *Intensive Animal Production in Developing Countries.* Occasional Publication 4—British Society of Animal Production. pp. 151–188.

INGRAM, W. M. and PRESCOTT, G. W. (1954). Toxic fresh-water algae. *Amer. Midl. Nat., 52 (1),* 75–87.

JACKSON, A. J., CAPPER, B. S. and MATTY, A. J. (1982). Evaluation of some plant proteins in complete diets for the tilapia *Saratherodon mossambicus. Aquaculture, 27,* 97–109.

JHINGRAN, V. G. and SHARMA, B. K. (1980). Integrated livestock-fish farming in India. In: R. S. V. Pullin and Z. H. Shehadeh (eds), *Integrated Agriculture-Aquaculture Farming Systems. ICLARM Conf. Proc., 4,* 135–142.

JOHN, C. K. (1975). Cultivation of fresh-water fish in the effluent pond of rubber processing factories. In: *Proceedings of the Agro-Industrial Wastes Symposium, Kuala Lumpur, Malaysia,* pp. 182–190.

JOLY, L. G. (1981). Feeding and trapping fish with *Piper auritum. Econ. Bot., 35 (4),* 383–390.

JUAREZ, A. (1976). Goat production in Mexico. Proceedings of a Workshop on the Role of Sheep and Goats in Agricultural Development. Winrock International Center, Arcansas.

KERNS, C. L. and ROELOFS, E. W. (1977). Poultry wastes in the diet of Israeli carp. *Bamidgeh, 29 (4),* 125–135.

KETOLA, H. G. (1982). Effects of phosphorous in trout diets on water pollution. *Salmonid, 6 (2),* 12–15.

KHOO, K. H. and TAN, E. S. P. (1980). Review of rice-fish culture in South East Asia. In: R. S. V. Pullin and Z. H. Shehadeh (eds), *Integrated Agriculture-Aquaculture Farming Systems. ICLARM Conf. Proc., 4,* 1–14.

KOEMAN, J. H. (1974). Problems associated with the use of agricultural chemicals. Conference on Ecological Guidelines for Forest, Land, Water Resources Development in Indonesia. University of Padjadjaran, Bandung, Indonesia. 11 pp.

KOESOEMADINATA, S. (1980). Pesticides as a major constraint to integrated agriculture-aquaculture farming systems. In: R. S. V. Pullin and Z. H. Shehadeh (eds), *Integrated Agriculture-Aquaculture Farming Systems. ICLARM Conf. Proc., 4,* 45–51.

KURONUMA, K. (1980). Carp culture in Japanese rice fields. In: R. S. V. Pullin and Z. H. Shehadeh (eds), *Integrated Agriculture-Aquaculture Farming Systems. ICLARM Conf. Proc., 4,* 167–174.

LAHSER, C. W. (1967). *Tilapia mossambica* as a fish for aquatic weed control. *Prog. Fish-Cult., 29,* 48–50.

LAM, Y. C. (1973). Common fertilisers in Hong Kong and their application methods. *Hong Kong Farmer, 5 (4),* 16–18.

LAWSON, S. (1981). *Inland Fish Culture in Thailand; a Case Study.* Systems Group, Open University, Milton Keynes, UK.

LEACH, G. (1975). *Energy and Food Production.* International Institute for Environment and Development, London.

LEACH, G. (1976). *Energy and Food Production.* IPC Business Press Ltd, Guildford, UK. 137 pp.

LE MARE, D. W. (1952). Pig rearing, fish farming and vegetable growing. *Malay. Agric. J., 35,* 156–166.

LEWIS, W. M. and BENDER, M. (1961). Effect of a cover of duckweeds and the alga *Pithophora* upon the dissolved oxygen and free carbon dioxide of small ponds. *Ecology, 42,* 602–603.

LIAO, I. C. and CHEN, T. P. (1983). Status and prospects of tilapia culture in Taiwan. In: L. Fishelson and Z. Yaron (eds), *Proceedings of International Symposium on Tilapia in Aquaculture,* Tel Aviv, Israel. pp. 588–596.

LING, S. W. (1967). Feeds and feeding of warm-water fishes in ponds in Asia and the Far East. *FAO Fish. Rep., 44 (3),* 291–309.

LINN, J. G., GOODRICH, R. D., OTTERBY, D. E., MEISKE, J. C. and STABA, E. J. (1975). Nutritive value of dried or ensiled aquatic plants. 2: Digestibility by sheep. *J. Ameri. Sci., 41,* 610–615.

LITTLE, D. C. (1983). An energy and nutrient study of integrated aquaculture in Thailand. M.Sc. thesis. University of Stirling, Scotland. 154 pp.

LITTLE, E. C. S. and HENSON, I. E. (1967). The water content of some important tropical water weeds. *Pest Artic. News Sum., 13,* 223–227.

LOVELL, T. (1979). Fish feed and nutrition. *Commer. Fish Farmer Aquacult. News, 5 (4),* 31–32.

LU, J. and KEVERN, N. (1975). The feasibility of using waste materials as supplemental fish feed. *Prog. Fish-Cult., 37,* 241–244.

LYAYMAN, E. M. (1966). Textbook on diseases of fish. IZD. Vysshaya Shkola, Moscow, 3, 115–122.

MA, A. M. (1975). Agro-industrial waste problems in Malaysia. *Proceedings of the Agro-Industrial Wastes Symposium,* Kuala Lumpur, Malaysia, pp. 1–7.

MALHERBE, H. H. and COETZEE, O. J. (1965). The survival of type '2' poliovirus in a model system of stabilisation ponds. *CSIR Research Report, 242,* Pretoria: Council for Scientific and Industrial Research.

MARA, D. D. (1977). Wastewater treatment in hot climates. In: R. Feacham, M. G. McGarry and D. Mara (eds), *Water, Wastes and Health in Hot Climates.* John Wiley and Sons, N.Y., USA, pp. 264–283.

MARTYSHEV, F. G. (1983). *Pond Fisheries.* A. A. Balkema, Rotterdam.

MATHES, H. (1978). The problem of rice-eating fish in the central Niger Delta, Mali. (Symposium on river and flood plain fisheries in Africa). *CIFA Tech. Pap., 5,* 225–252.

MCGARRY, M. G. (1976). The taboo resource, the use of human excreta in Chinese agriculture. *Ecologist, 6 (4).*

MCGARRY, M. G. (1977[a]). Domestic wastes as an economic resource: biogas and fish culture. In: R. Feacham, M. G. McGarry and D. D. Mara (eds), *Wastes and Health in Hot Climates.* John Wiley and Sons, N.Y., USA, pp. 347–364.

MCGARRY, M. G. (1977[b]). Waste collection in hot climates: a technical and economic appraisal. In: R. Feacham, M. G. McGarry and D. Mara (eds), *Water Wastes and Health in Hot Climates.* John Wiley and Sons, N.Y., USA, pp. 239–263.

MCGARRY, M. G. and TONGKAME, C. (1971). Water reclamation and algae harvesting. *J. Water Pollut. Control Fed., 43,* 824–835.

MCLARNEY, B. (1978). Family-scale aquaculture in Columbia. Professor develps the peasant fish culture unit. *Commer, Fish Farmer Aquacult. News, 4* (6), 28-29.

MCLARNEY, W. O., HENDERSON, S. and SHERMAN, M. M. (1974). A new method for culturing *Chironomus tentans.* Fabricius larvae using burlap substrate in fertilised pools. *Aquaculture, 4* (3), 267-276.

MCLARNEY, W. O., HENDERSON, S. and SHERMAN, M. M. (1977). Midge (*Chironomid*) larvae as a growth promoting supplement in fish and lobster diets. *Bamidgeh, 29.*

MEHTA, I., SHARMA, R. K. and TUANK, A. P. (1976). The aquatic weed problem in the Chambal irrigated area and its control by grass carp. In: C. K. Varshney and J. Rzoska (eds), *Aquatic weeds in South East Asia.* D. & W. Junk, The Hague, Netherlands, pp. 307–314.

MEYNELL, P. J. (1982). *Methane: Planning a Digester.* (2nd edition), Prism Press, UK.

MICHAEL, A. M. (1978). *Irrigation Theory and Practices.* Vikas Publishing House Private Limited, New Delhi, India.

MIDDENDORP, A. J. and VERRETH, J. A. J. (1986). The potential of and constraints to fish culture in integrated farming systems in the Lam Pao Irrigation Project, Northeast Thailand. *Aquaculture, 56,* 63–78.

MILLER, B. F. and SHAW, J. H. (1969). Digestion of poultry manure by Diptera. *Poult. Sci., 48 (5),* 1944–1945.

MIRES, D. (1983). The development of freshwater prawn *Machrobrachium rosenbergii* culture in Israel. *Bamidgeh, 35 (3),* 63–72.

MOAV, R. G., WOHLFARTH, G. J., SCHROEDER, G., HULATA and BARASCH, H. (1977). Intensive polyculture of fish in freshwater ponds. I: substitution of expensive feeds by liquid manure. *Aquaculture, 10,* 25–43.

MORGAN REES, A. M., WILLIAMS, T. E., SMITH, A. J. and CAPPER, B. S. (1977). Report of ODA Mission to Egypt to undertake a prefeasibility study of forage production and animal feeds. Tropical Products Institute, London.

MORIARTY, D. J. W. (1973). The physiology of digestion of blue green algae in the cichlid fish *Tilapia nilotica. J. Zool. (Lond.), 171,* 25–40.

MORIARTY, D. J. W. and MORIARTY, C. M. (1973). The assimilation of carbon from phytoplankton by two herbivorous fishes. *Tilapia nilotica* and *Hapochromis nigripinnis*. *J. Zool. (Lond.), 171,* 41–55.

Morrison, F. (1959). *Feeds and Feeding*. Monison Publ. Co., Iowa, USA.

Muir, J. P. and Beveridge, M. C. M. (1987). Water resources and aquacultural development. Proceedings of International Conference on Limnology. Kuala Lumpur, Malaysia, December, 1982. *Arch. Hydrobiol.* (in press).

Muller, R. (1978). The aquacultural rotation. *Aquaculture Hungarica (Szawas), 1,* 73–79.

Muller, Z. O. (1982). Feed from animal wastes: feeding manual. *FAO Animal Production Health Paper, 28,* 214 pp.

Myers, R. W. (1977). A comparative study of nutrient composition and growth by selected duckweeds, Lemnaceae, on dairy waste lagoons. M.Sc. thesis, Louisiana State University, 74 pp.

Neuhauser, E. F., Harteinstein and Kaplan, D. L. (1980). Growth of the earthworm *Eisenia foetida* in relation to population density and food rationing. *Oikos, 35,* 93–98.

Nimpuno, K. (1981). The Vietnamese toilet. *Appropriate Technology, 7 (4),* 15–17.

Nugent, G. G. (1978). Integration of the husbandry of farm animals and fish with particular reference to pig raising in tropical areas. In: C. M. R. Pastakia (ed), *Report of the Proceedings—Fish Farming and Wastes.* Institute of Fish Management and Society of Chemical Industries, London, UK.

Odum, E. P. (1975). *Ecology. The Link between the Natural and the Social sciences.* University of Georgia, USA.

Odum, E. P. and De La Cruz, A. (1967). Particulate organic detritus in Georgia salt marsh-estuarine ecosystem. *Estuaries, 83,* 383–389.

Panayotou, T. (1982). Social welfare economics and aquaculture. Issues for policy and research. In: *Aquaculture Economics Research in Asia:* Proceedings of a workshop held in Singapore 2–5 June 1981, Ottawa, Ontario, IDRC. 128 pp.

Pantalu, V. R. (1980). Aquaculture in irrigation systems. In: R. S. V. Pullin and Z. H. Shehadeh (eds), *Integrated Agriculture-Aquaculture Farming Systems. ICLARM Conf. Proc., 4,* 35–44.

Pantastico, J. B. and Baldia, J. P. (1979). Supplemental feeding of *Tilapia mossambicus.* In: J. E. Halver and S. K. Tiew (eds), *Finfish Nutrition and Fish Feed Technology.* Vol. 1. Heinemann Verlagsgessellschaft mGH, Berlin, pp. 587–593.

Pantastico, J. B. and Baldia, J. P. (1980). Ipil-ipil leaf meal as supplement feed for *T. nilotica* in cages. *Fish. Res. J. Philipp., 5 (2),* 63–68.

Parr, J. F. Epstein, E. and Willson, G. B. (1978). Composting sewage sludge for bed application. Agric. Environ., *4* (2), 123-138.

Payne, W. J. A. (1981). The desirability and implications of encouraging intensive animal production enterprises in developing countries. In: A. J. Smith and R. G. Gunn (eds), *Intensive Animal Production in Developing Countries. Occassional Publication 4—British Society of Animal Production,* pp. 1–10.

Philippart, J. C., Melard, C. H. and Runet, J. G. (1979). La pisciculture dans les effluents thermiques industriels. Bilan et perspectives d'une annee de recherche a la centriale nucleaire de Tihange Sur la Meuse. In: L. Calembert (ed), *Problematique et Gestion des Eaux Interceives.* pp. 779–791.

Pimentel, D. *et al.* (1973). Food production and the energy crisis. *Science, 182,* 443.

Pimentel, D., Beraredi, G. and Fast, S. (1983). Energy efficiency of farming systems: organic and conventional agriculture. *Agric. Ecosyst. and Environ., 9,* 359–372.

Plavnik, I., Barasch, H. and Schroeder, G. (1983). Utilisation of ducks droppings in fish farming. *Nutr. Rep. Int., 28,* 3.

Plucknett, D. L. and Beemer, H. (1981). Vegetable farming systems in People's Republic of China. Westview Special Studies in Agricultural Science and Policy, 386 pp.

Polich, T. (1979). Rice/carp farming in the Philippines and cultural acceptance. *ICLARM Newsl., 2 (4),* 13–15.

Pongsuwana, U. (1963). Country statement—Thailand. In: *Report of Association of South East Asian Nations.* ASEAN Seminar on Shrimp Culture, 3 pp.

Popma, T. J. (1978). Ensayo sobre el crecimiento de *Tilapia rendalli* Boulenger, enjaulada y alimentado con follaje de bore (*Alocasia macrorhiza*). *Cent. Piscic. Exp. Int. Téc., 2,* 43–50.

Porath, D., Hepher, B. and Koton, A. (1979). Duckweed as an aquatic crop evaluation of clones for aquaculture. *Aquat. Bot., 7 (3),* 273–278.

Pradt, L. A. (1971). Some recent developments in night soil treatment. *Water Res., 5,* 507–521.

Probst. (1934). Teichwirtschaft und Geflugelzucht in ihren Wechsellbeziehungen (Pond culture, duck raising and their interactions). *Handbuch der Binnenfischerei Mitteleuropas, 4,* 408–82.

PULLIN, R. S. V. (1982). Experimental integrated farming systems in Mexico. *ICLARM Newsl., 5,* 11.

PULLIN, R. S. V. (1983). Culture of herbivorous tilapias. Paper presented at International Conference on Development and Management of Tropical Living Aquatic Resources. August 2–5, 1983. Universiti Pertanian, Malaysia.

PULLIN, R. S. V. (1985). Time to reappraise rice fish culture. *ICLARM Newsl., 8 (4),* 3–4.

PULLIN, R. S. V. and ALMAZAN, G. (1983). *Azolla* as a fish feed. *ICLARM Newsl., 6 (1),* 6–7.

RAA, J. and GULBERG, A. (1982). Fish silage: a review. *CRC Crit. Rev. Food Sci. Nutr.,* April 1982, pp. 383–419.

RABANAL, H. R. and SHANG, Y. C. (1979). The economics of various management techniques for pond culture of finfish. In: T. V. R. Pillay and W. A. Dill (eds), *Advances in Aquaculture.* Fishing News Books Ltd, Farnham, UK. pp. 224–235.

RAMOS-HENAO, A. and CORREDOR, G. G. (1980). Effects of three management practices on the growth and production of *T. rendalli. Bamidgeh, 32,* 41–45.

RAMSEY, P. (1983). Rice-fish practices in Ifagao Province, Philippines. *ICLARM Newsl., 6 (3),* 8.

RAPPAPORT, U. and SARIG, S. (1978). The results of manuring on intensive growth fish farming at the Ginosar station ponds in 1977. *Bamidgeh, 30 (2),* 27–36.

REICH, K. (1975). Multispecies fish culture (polyculture) in Israel. *Bamidgeh, 27,* 85–99.

RIFAI, S. A. (1979). The use of aquatic plants as feed for *Tilapia nilotica* in floating cages. In: *Proceedings of International Workshop on Pond Cage Culture of Fish.* International Development Research Centre, Canada, Aquaculture Dept. South East Asian Fisheries Development Centre, Iloilo, Philippines. pp. 61–64.

RIFAI, S. A. (1980). Control of reproduction of *Tilapia nilotica* using cage culture. *Aquaculture, 20,* 177–185.

RIJN, VAN J., DIA, S. and SHILO, J. M. (1984). Mechanisms of ammonia transformation in fish ponds. In: H. Rosenthal and S. Saig (eds), *Proceedings of the Second Seminar of the German-Israeli Co-operation in Aquaculture Research. Special Publication 8, European Mariculture Society.* pp. 17–40.

ROY-SMITH, F. (1981). Intensive milk production from goats in developing countries. In: A. J. Smith and R. G. Gunn (eds), *Intensive Animal Production in Developing Countries. Occasional Publication 4—British Society of Animal Production,* pp. 387–394.

RUDDLE, K. (1985[a]). Labor supply and demand in a complex system: Integrated agriculture-aquaculture in the Zhujiang Delta, China. *Bulletin of the National Museum of Ethnology, 10 (3),* 773–819.

RUDDLE, K. (1985[b]). Rural reforms and household economics in the dike-pond area of the Zhujiang Delta, China. *Bulletin of the National Museum of Ethnology, 10 (4),* 1145–1174.

RUDDLE, K., FURTADO, J. I., ZHONG, G. F. and DENG, H. Z. (1983). The mulberry dike-carp pond resource system of the Zhujiang (Pearl River) Delta, People's Republic of China. 1: Environmental context and system overview. *Applied Geography, 3,* 45–62.

RUIZ, M. E. (1981). The use of green bananas and tropical crop residues for intensive beef production. In: A. J. Smith and R. G. Gunn (eds), *Intensive Animal Production in Developing Countries. Occasional Publication 4—British Society of Animal Production.* pp. 371–383.

RUMICH, B. (1967). *The Goats of Indonesia.* FAO Regional Office, Bangkok, Thailand.

RUSKIN, F. R. and SHIPLEY, D. W. (1976). *Making Aquatic Weeds Useful: Some Perspectives for Developing Countries.* National Academic of Sciences, Washington, D.C., USA.

RYBEZYNSKI, W. C., POLPRASERT, W. C. and MCGARRY, M. (1978). Appropriate waste treatment for tropical climates: a state-of-the-art review. In: E. A. R. Ouano, B. N. Lohani and N. C. Thanh (eds), *International Conference on Water Pollution Control in Developing Countries.* Bangkok, Thailand. pp. 735–742.

SARIG, S. (1971). The prevention and treatment of diseases of warm water fishes under sub-tropical conditions, with special emphasis on intensive fish farming. In: S. F. Snieszko and H. R. Axelood (eds), *Diseases of Fishes, Book 3.* T. F. H. Publications, Hong Kong.

SARIG, S. (1984). The integration of fish culture into general farm irrigation systems in Israel. *Bamidgeh, 1 (36),* 16–20.

SCHROEDER, G. L. (1974). Use of fluid cowshed manure in fish ponds. *Bamidgeh, 26 (3),* 84–96.

SCHROEDER, G. L. (1975). Some effects of stocking fish in waste treatment ponds. *Water Res., 9,* 591–593.

SCHROEDER, G. L. (1978). Autotrophic and hetertrophic production of micro-organisms in intensively-manured fish ponds and related fish yields. *Aquaculture, 14,* 303–325.

SCHROEDER, G. L. (1980). Fish farming in manure-loaded ponds. In: R. S. V. Pullin and Z. H.

Shehadeh (eds), *Integrated Agriculture-Aquaculture Farming Systems. ICLARM Conf. Proc. 4,* 73–86.

SCHROEDER, G. L. (1983). Sources of fish and prawn growth in polyculture ponds as indicated by S.C. analysis. *Aquaculture, 35,* 29–42.

SCHROEDER, G. L. and HEPHER, B. (1979). Use of agricultural and urban wastes in fish culture. In: T. V. R. Pillay and W. A. Dill (eds), *Advances in Aquaculture,* Fishing News Books Ltd, Farnham, UK, pp. 487–489.

SCHUSTER, W. H. (1952). Fish culture as a means of controlling aquatic weeds in inland waters. *FAO Fish. Bull., 5 (1),* 15–24.

SHANG, Y. C. (1981). *Aquaculture Economics: Basic Concepts and Methods of Analysis.* Croom Helm, London, UK.

SHAW, P. C. and MARK, K. K. (1980). Chironomid farming—a means of recycling farm manure and potentially reducing water pollution in Hong Kong. *Aquaculture, 21 (2),* 155–163.

SHELEF, MORRINE AND OZON (1978). Animal feed protein and water for irrigation from algal ponds. In: E. A. R. Ovano, B. N. Lohani and N. C. Thanh (eds), *International Conference on Water Pollution Control in Developing Countries.* Bangkok, Thailand. pp. 217–228.

SIN, A. W. (1980). Integrated animal fish husbandry systems in Hong Kong, with case studies on duck-fish and goose-fish systems. In: R. S. V. Pullin and Z. H. Shehadeh (eds), *Integrated Agriculture-Aquaculture Farming Systems. ICLARM Conf. Proc., 4,* 113–123.

SINGH, V. P., EARLY, A. C. and WICKHAM, T. H. (1980). Rice agronomy in relation to rice-fish culture. In: R. S. V. Pullin and Z. H. Shehadeh (eds), *Integrated Agriculture-Aquaculture Farming Systems. ICLARM Conf. Proc., 4,* 15–34.

SMITH, A. (1979). The role of draught animals in agricultural systems in developing countries. In: C. R. W. Spedding (ed), *Vegetable Production.*

SOMMER, T. (1984). California rice fields have cray fish potential. *Aquaculture Magazine, 10 (5),* 6.

SPATURU, P. (1976). Natural feed of *Tilapia aurea* (Steindachiner) in polyculture, with supplementary feed and intensive manuring. *Bamidgeh, 28,* 57–63.

SPATURU, P. (1977). Gut contents of silver carp. *Hypophthalmichthys molitrix*—and some trophic relations to other fish species in a polyculture system. *Aquaculture, 11,* 137–146.

SPATURU, P. (1978). Food and feeding habits of *Tilapia zillii* (Gervais) (Cic lidae) in Lake Kinneret (Israel. *Aquaculture, 14,* 327–338.

SPILLER, G. (1985). *Rice cum Fish Culture.* Working Paper Fisheries Advisor Project DOF/CIDA (906/08201), Dept. of Fisheries, Bangkok, Thailand. 48 pp.

STAFFORD, E. M. (1984). The use of earthworms as a feed for rainbow trout, *Salmo gairdneri.* Ph.D. thesis, University of Stirling, Scotland, 178 pp.

STAFFORD, E. M. and TACON, G. J. (1984). Nutritive value of the earthworm *Dendrodrilus subrubicumdus* grown on domestic sewage in trout diets. *Agricultural Wastes, 9,* 249–266.

SUWANASART, P. (1972). Effects of feeding, mesh size and stocking size on the growth of *Tilapia aurea* in cages. *Annu. Rep. Int. Cent. Aquaculture, Auburn University, Auburn, Alabama,* 1971, 71–79.

SWAMINATHAN, M. S. (1984). Rice. *Scientific American, 250 (1),* 63–71.

SWICK, R. A., CHEEKE, P. R. and PATTON, N. M. (1978). Evaluation of dried rabbit manure as a feed for rabbits. *Can. J. Anim. Sci., 58,* 753–757.

SWINGLE, H. S. (1972). Relationship of the Thai-fish culture programme to production of fish in the lower Mekong area. SEADAG Mekong Development Seminar.

TACON, A. (1978). The use of activated sludge as a supplementary protein source in the diet of rainbow trout. Ph.D. thesis, University College, Cardiff, 215 pp.

TACON, A. (1981). The possible substitution of fish meal in fish diets. *Proceedings of the SMBA-HIDB Fish Farming Meeting,* Oban, Scotland, February 26–27(1981), pp. 45–56.

TACON, A. (1983). *Aquaculture in Bolivia. Evaluation of the Development Corporation for Chukuisaca Department (CORDECH) Tarija Department (CODETAR) and San Jacinto Association (ASJ-Tarija) Fish Farming Projects in Bolivia.* Report for the Overseas Development Administration, London. 120 pp.

TACON, A. G. J., STAFFORD, E. A. and EDWARDS, E. A. (1983). A preliminary investigation of the nutritive value of three terrestrial lumbricid worms for rainbow trout. *Aquaculture, 35,* 187–199.

TAIGANIDES, E. P. (1977). Bioengineering properties of feedlot wastes. In: E. P. Taiganides (ed), *Animal Wastes.* Applied Science Publ. Ltd, Essex, UK.

TAN, C. E., CHONG, B. J., SIER, H. K. and MOULTON T. (1973). A report on paddy and paddy field fish production in Krian, Perak. *Bulletin, 128,* Ministry of Agriculture and Fisheries, Malaysia. 58 pp.

TAN, E. S. P. and KHOO, K. K. (1980). The integration of fish farming with agriculture in Malaysia. In: R. S. V. Pullin and Z. H. Shehadeh (eds), *Integrated Agriculture-Aquaculture Farming Systems. ICLARM Conf. Proc., 4,* pp. 175–188.

TANG (1970). Evaluation of balance between fishes and available fish ponds in multi species fish culture ponds in Taiwan. *Trans. Am. Fish. Soc., 99,* 708–718.

TAPIADOR, D. D., HENDERSON, H. F., DELMENDO, M. N. and TSUTSUI, H. (1977). Freshwater fisheries and aquaculture in China. *FAO Fish. Tech. Pap., 168,* Rome. 83 pp.

TATTERSON, I. N. and WINDSOR, M. L. (1974). Fish silage. *Torry Advisory Note 64,* 6 pp. Ministry of Agriculture, Fisheries and Food, Torry Research Station, Aberdeen, Scotland.

TAYLOR, S. R. KLAMPRATUM, D. and NATENGUM, P. (1986). *The Synergistic Effect of Border-Effect and Fish on a Rice-Fish Strategy in Northern Thailand.* Report to Farming Systems Research Institute, Dept. of Agriculture, Chiang Mai, Thailand. 13 pp.

VAAS, K. F. and SACHLAN, M. (1956). Cultivation of common carp in running water in West Java. *Proceedings of the 6th Session IPFC Tokyo,* 1955, FAO, Bangkok, Thailand.

VAN DYKE, J. M. and SUTTON, D. L. (1977). Digestion of duckweed (*Lemna spp.*) by the grass carp (*Ctenopharyngodon idella*). *J. Fish Biol., 11,* 273–278.

VAN DER LINGEN, M. I. (1968). Control of pond weeds. *FAO Fish. Rep., 44 (5),* 53–60.

VAN WEERD, J. H. (1985). Growth and survival in drainage channels of grass carp. *Ctenopharyngodon idella* Val, fry and their potential for weed control. *Aquacult. and Fish. Manage., 16,* 7–24.

VARDI, Y. (1980). National water resources planning and development in Israel—the endangered resource. In: H. I. Shuval (ed), *Water Quality Management under Conditions of Scarcity—Israel as a Case Study.* Academic Press, N.Y., USA, pp. 37–51.

VELASQUEZ, C. C. (1980). Health constraints to integrated animal fish farming in the Philippines. In: R. S. V. Pullin and Z. H. Shehadeh (eds), *Integrated Agriculture-Aquaculture Farming Systems. ICLARM Conf. Proc., 4,* 103–111.

VIOLA, S. and ARIELI, Y. (1983). Nutrition studies with tilapia (*Saratherodon*) 1: Replacement of fish meal by soybean meal in feeds for intensive tilapia culture. *Bamidgeh, 35 (1),* 9–17.

VIOLA, S., MOKADI, S., RAPPAPORT, U. and ARIELI, Y. (1982). Partial and complete replacement of fish meal by soybean meal in feeds for intensive culture of carp. *Aquaculture, 26,* 223–236.

WANG, S. (1985). Nutritive value of leucaena leaf meal in pelleted feed for tilapia. M.Sc. thesis, Asian Institute of Technology, Bangkok, Thailand. 50 pp.

WATT, W. L. (1942). Stall-feeding of goats and sheep by the Kikya Tribe, Kenya. *East Afr. Agric. J., 8,* 109–111.

WEE, K. L., KERDCHUEN, N. and EDWARDS, P. (1986). Use of waste grown tilapia silage as feed for *Clarias betrachus* (Linn.). Paper presented at the First Asian Fisheries Forum, Manila, 25–31 May 1986.

WELCOMME, R. L. (1979). Extensive aquaculture practices on African floodplains. *CIFI Tech. Pap., 4,* 248–252. FAO, Rome.

WESTLAKE, D. F. (1963). Comparisons of plant production. *Biol. Rev., 38,* 385–425.

WETCHARAGARUM, K. (1980). Integrated agriculture-aquaculture studies in Thailand, with a case study on chicken-fish farming. In: R. S. V. Pullin and Z. H. Shehadeh (eds), *Integrated Agriculture-Aquaculture Farming Systems. ICLARM Conf. Proc., 4,* 243–249.

WOHLFARTH, G. W. and SCHROEDER, G. L. (1979). Use of manure in fish farming—a review. *Agricultural Wastes, 1,* 279–299.

WOLVERTON, S. C. and MCDONALD, R. C. (1979). The water hyacinth from prolific pest to potential provider. *Ambio., 8,* 2–9.

WOYNAROVICH, E. (1979). The feasibility of combining animal husbandry with fish farming with special reference to duck and pig production. In: T. V. R. Pillay and W. A. Dill (eds), *Advances in Aquaculture.* Fishing News Books Ltd, Farnham, UK. pp. 203–208.

WOYNAROVICH, E. (1980). Raising ducks on fish ponds. In: R. S. V. Pullin and Z. H. Shehadeh (eds), *Integrated Agriculture-Aquaculture Farming Systems. ICLARM Conf. Proc., 4,* 129–134.

WOYNAROVICH, E. and KUNHOLD W. W. (1979). Report of consultancy to Penang, Malaysia regarding animal waste management problem. *FAO-S.C.S./79/WP/86,* 59 pp.

YAKUPITIYAGE, A. (1984). A feasibility study of the culture of earthworms on cesspool slurry and the evaluation of their nutritive value for the catfish *Clarias macrocephalus.* M.Sc. thesis, Asian Institute of Technology, Bangkok, Thailand. 97 pp.

YAP, T. N., GHAZALI, M., PHANG, S. M. and RAHMAN, R. A. (1983). Potential of rubber effluent treatment ponds for aquaculture. Paper given at International Conference on Development and Management of Tropical Living Aquatic Resources. Universiti Pertanian Malaysia, August 2–5, 1983.

YASHOUV, A. (1970). Propagation of chironomid larvae as food for fish fry. *Bamidgeh, 22 (4),* 101–105.

YASHOUV, A. (1971). Interaction between common carp and silver carp in fish ponds. *Bamidgeh, 23,* 85–92.

YOUNT, J. L. and CROSSMAN, R. A. (1970). Eutrophication control by plant harvesting. *J. Water Pollut. Contr. Fed., 42,* R173–183.

ZHU DE SHAN (1980). A brief introduction to the fisheries of China. *FAO Fish. Circ., (726),* 3 pp.

ZONNEVELD, N, and VAN ZON, H. (1985). The biology and culture of grass carp (*Ctenopharyngodon idella*) with special reference to their utilisation for weed control. In: J. F. Muir and R. J. Roberts (eds), *Recent Advances in Aquaculture Vol. 2,* Croom Helm, London, UK.

ZWEIG, R. D. (1984[a]). Aquaculture strategies in China. *Oceanus, 26,* (4), 33–39.

ZWEIG, R. D. (1984[b]). Culture of Chinese carp fry using Barnyard Grass. *Aquaculture Magazine,* 16.

INDEX TO GENERA AND SPECIES

GENERAL INDEX